Making Peace with the Earth

Making Peace with the Earth

Action and Advocacy for Climate Justice

Edited by Grace Ji-Sun Kim

Foreword by Guillermo Kerber

World Council of Churches Publications

MAKING PEACE WITH THE EARTH
Action and Advocacy for Climate Justice
Edited by Grace Ji-Sun Kim

WCC Publications is the book publishing programme of the World Council of Churches. Founded in 1948, the WCC promotes Christian unity in faith, witness and service for a just and peaceful world. A global fellowship, the WCC brings together 345 Protestant, Orthodox, Anglican and other churches representing more than 550 million Christians in 110 countries and works cooperatively with the Roman Catholic Church.

Opinions expressed in WCC Publications are those of the authors.

Cover design, book design and typesetting: Michelle Cook / 4 Seasons Book Design
Cover image: Lucy D'Souza-Krone, "God's Eye Watches over the Earth." This image and others on pages 182-86 are used by permission of the artist. See www.Lucy-art.de and contact at Andreas.Lucy@t-online.de.

ISBN: 978-2-8254-1668-6

World Council of Churches
150 route de Ferney, P.O. Box 2100
1211 Geneva 2, Switzerland
http://publications.oikoumene.org

CONTENTS

FOREWORD

Guillermo Kerber

This volume draws on the lived experience of Christian churches around the world as they address the problems, perils, and possibilities inherent in climate change.

In 2013 the World Council of Churches Assembly in Busan, Republic of Korea, concluded by inviting Christians and all people to join a pilgrimage of justice and peace. In 2015, in various countries, pilgrimages for climate justice were organized by churches, specialized ministries, and ecumenical and civil society organizations. Many of these pilgrimages were the result of what churches have been doing on climate and ecological issues for many years. Climate justice is a complementary dimension of peace with the earth, one of the components of the International Ecumenical Peace Convocation which concluded the Decade to Overcome Violence, Churches Seeking Reconciliation and Peace, held in Kingston, Jamaica, in 2011.

In recent years, more and more publications in various regions and in different languages are addressing the climate and ecological crises from theological perspectives. These theological reflections have been inspired by concrete actions been taken by churches on the ground. At the same time the theological insights have inspired Christians to translate into action a renewed concern for creation.

The purpose of this book it to show through some particular cases how churches and church-related organizations have been translating into practice, through action and advocacy, peace with the earth and climate justice. Together with concrete actions being taken by churches on the ground (green churches and eco-congregations being an example of these), churches' advocacy at the national and international level on environmental issues has a history of more than twenty years. The Earth Summit or United Nations Conference on Environment and Development (UNCED), which took place in Rio de Janeiro, Brazil, in 1992, can be seen as a catalyst in the international arena as three environmental conventions were adopted there by the international community: the convention on biological diversity, the convention on desertification, and the convention on climate change. Churches, as we will see in one of the chapters of this book, were strongly involved in Rio 1992 and since then, especially at the Conference of Parties (COP) of the United Nations Convention on Climate Change that started in 1995.

The World Council of Churches' Working Group on Climate Change, also in existence for more than twenty years, has been a space where representatives of member churches, other churches, specialized ministries and civil society organizations working on care for creation and climate justice have come together to share experiences, learn from each other and strategize for advocacy at COPs and at the United Nations Human Rights Council. Most of the chapters of this book have been written by members of the Working Group. But, because the group does not cover the full variety of actions being developed on these issues, other people were also invited to contribute to this volume.

Despite this long history of ecumenical engagement, in many places ecology and specifically climate change continue to be a topic far from the action of the churches. Publication of this book is meant to share with a wider public stories and reflections from very diverse contexts and with very diverse styles. The contributions reflect the richness of the ecumenical movement in addressing the challenges and also the difficulty in finding a common pattern. Yet "caring for creation" and "striving for climate justice" can be seen as concepts that reflect what churches have been doing based on a theology of creation which responds to the current ecological crisis and an ethics of climate that recognizes that those countries and communities who suffer today (and will suffer tomorrow) the worst consequences of climate change are those who have contributed the least to the causes, mainly the increase in CO_2 emissions.

The ecumenical work on the ecological and climate crises has been done not in isolation but in close cooperation with other churches, other faith traditions and the global civil society movement.

Pope Francis's encyclical *Laudato si* shows how the topics that have been at the core of the ecumenical movement's work for climate justice are also essential to the social teaching of the Roman Catholic Church. Care for creation as our common home, the need to listen to the cry of the earth together with the cry of the poor, the need for profound change from the individual to the international level, a deep spirituality of creation and the relevance of education are some of the key common aspects highlighted by the encyclical.

Recent interfaith statements, like those adopted in New York in September 2014 at the Climate Change Summit or presented to the UNFCCC Secretariat in October 2015, reveal an increasing interfaith consensus in addressing the topic. The statements, based on the different faith traditions, called the international community to agree on a fair, ambitious, and binding treaty to respond to the climate change threat, and express the commitment of faith leaders to implement in their communities the necessary changes.

Campaigns and advocacy at the international level have also been undertaken by the ecumenical movement in cooperation with larger non-governmental organizations and networks. The World Council of Churches has worked especially with the Global Call for Climate Action, linking faith voices to other voices which share the same perspectives and concerns, and with 350.org in mobilizing people all over the world for an effective response to the threats humanity is facing because of climate change. The call to divest from fossil fuels, which has been listened to by many churches, theological seminaries, and other church-related organizations, with the WCC's Central Committee having explicitly included fossil fuels among the list of areas where the WCC will not invest, is one example of this collaboration.

As a result of long years of negotiations and strong advocacy from faith communities and civil society at large, the 21st meeting of the Conference of Parties of the United Nations Framework Convention on Climate Change adopted in December 2015 the so called Paris Agreement. The international community has reached a consensus on how to address the climate crisis, which is very much welcomed, but much more is needed to hold "the increase in the global average temperature to well below 2°C above pre-industrial levels and to pursue efforts to limit the temperature increase to 1.5°C," as the Agreement expresses in Article 2. Courageous decisions will have to be taken by governments to limit the increase in the average global temperature. Actions and advocacy by churches and faith-based organizations cannot stop now but should look for other creative forms of engagement to inspire and put into practice the needed "ecological conversion," as Pope Francis put it, for the care of our common home.

While sharing a host of examples of churches' action and reflection on the topic, this book will, we hope, also be an invitation to further commit and share stories of action and advocacy for climate justice.

Guillermo Kerber
Programme on Care for Creation and Climate Justice
World Council of Churches

PREFACE

Grace Ji-Sun Kim

I remember visiting my grandmother's house when I was a young girl living in Korea.[1] I have fond memories of playing in the *madang* (a Korean term that describes a courtyard within a traditional Korean home). Most Korean houses are surrounded by a large metal fence with a front gate that leads into the *madang*.

A typical part of Korean homes, the *madang* is much like a family room where family and friends gather to rest, talk, share, and engage with each other. Most of the rooms in the home are entered through the *madang* and thus it serves as a space for encounter and sharing, celebration and fellowship, greeting a visitor and welcoming a stranger. Ancient, medieval, and even contemporary European buildings are built on the same principle. Perhaps the *madang* is something like the courtyard at the Cloisters in New York City, although the surrounding building there is far larger than what my grandmother had or what most Korean homes would have. My grandmother's home was small, with just two rooms leading away from the *madang*.

The tenth assembly of the World Council of Churches (WCC) was held in Busan, South Korea, from 30 October to 8 November 2013. Participating churches, along with ecumenical organizations and groups, created a *madang* with information booths, cultural performers, and workshops. This *madang* became an energetic meeting place where individuals and groups continued ongoing conversations, shared, brainstormed, envisioned, and reimagined.

The participants in the tenth assembly adopted a message that asked churches and Christians to join in a "Pilgrimage of Justice and Peace." Many are now working on what it means to be on this pilgrimage.

Part of this journey and pilgrimage involves doing environmental justice and making peace with the earth. And this requires working on the pressing issue of climate change, one of the most challenging realities in today's world. People and vulnerable groups—in particular, communities—face climate change in various parts of the world. Churches and other faith-based communities are called to address climate change. Some Christians and churches fail to recognize the reality of climate change, however, because of their views on the Bible and faith. Others do not believe in climate change because of culturally conditioned climate skepticism. Yet others believe that it is not the role of faith communities to address climate change, feeling that it is not religion's job to engage any political issues.

Why should churches address environmental or climate-change issues? Why should Protestants join Orthodox, Roman Catholics, and Evangelicals in caring for creation initiatives? Why should Christians look to join interfaith work with Jews, Muslims, Hindus, Buddhists, and other faith traditions? Why should they advocate for climate justice and peace with the earth? These are just some of the questions that Christians, churches, and faith communities face when addressing climate change.

In light of these questions and doubts, the WCC is addressing climate change for two specific reasons. First, churches in various parts of the world recognize how the consequences of climate change affect the lives and livelihoods of their communities. In an effort to respond to these consequences, churches are helping to develop resilient communities, equipped to adapt to climate change. These churches benefit greatly by learning from the work of other faith communities from around the world. Second, at the local level, churches of different denominations are coming together to respond to the impacts of climate change and to advocate at local and national levels for policies that respond to the needs and rights of vulnerable populations. In both instances, the WCC has a unique ability to build networks and relationships between churches around the globe, and to accompany and enhance their work.

As scientific reports confirm, climate change affects the people who have been made the most vulnerable in our societies. Climate change intertwines with issues of land, food, work, devastation, and human dignity. Climate change often results in people becoming vulnerable to losing their land and being forced to live in other areas that are not devastated by climate change. The rights of the victims of climate change are being violated in various ways and the threats to human beings and to life in our planet will only get worse.

Religious leaders and faith-based organizations must embrace the climate-change issue vigorously. Advocacy for the earth must become a priority. For example, all must work towards preventing Earth's temperature from rising more than 2 degrees Celsius. Unlimited carbon pollution must be stopped. There needs to be a strong mitigation of CO_2 emissions by wealthy countries, so those countries still developing do not suffer from a burnt-out planet, through no fault of their own. All these steps are crucial to address climate change.

Environmental justice is intimately related to economic justice. The environment affects our economy, and people who live in poverty are the most affected. To emphasize and illustrate the link between these two intertwined issues, the WCC adopted the term *eco-justice*.

Furthermore, without environmental justice, there cannot be peace. The WCC's International Ecumenical Peace Convocation, held in Kingston, Jamaica, in 2011 clearly stated that there can be "*no peace on earth without peace with the earth*." The way we have related to the earth, however, is with little conservation and no restraint. We have lost any tradition of being stewards of the earth. This path of violence and domination may not lead to open war, but it contributes to the destruction of the earth.

The earth cries out and pleads with us to join together to work toward climate justice and environmental justice. We need to advocate for the earth and be at *peace with the earth*. We need to come into the earth's *madang* and be in that space to share, communicate, fellowship, greet a stranger, and welcome a friend. We need to bring Christians, Hindus, Buddhists, Muslims, individuals, communities, politicians, churches, and corporations into the *madang*, so all can challenge each other to live a life of stewardship rather than of greed, domination, and destruction. We all need to turn to the God of life and ask God to lead us to social justice and peace.

To respond to this threat to humanity and to earth as a whole, even climate-change skeptics must join in the journey for the protection and sustainability of the earth. This book tackles the important ecological-justice questions of our time. It is a collection from climate activists, faith leaders, theologians, and others who are on the pilgrimage of justice and peace, reflecting on climate justice and peace with the earth. The concrete experience of churches around the world can be instructive for the reader's church, and this volume gathers that hard-earned knowledge. This book seeks to provide imaginative possibilities of how the church, faith communities, individuals, and the academy can move forward to help save God's creation. It serves as an example of a virtual *madang*. It is my hope that readers will create *madangs* around the world: linked-together places for the worldwide church to work together to make changes in the world that will help lead toward peace on earth. The *madang* is open; let us come in, converse, envision, and reimagine. The journey has already begun and our hope is that everyone will join in this journey of justice and peace with the earth.

Acknowledgments

This book rose out of the 2014 Climate Change Working Group for the World Council of Churches at Wuppertal, Germany. Through the various discussions, site visits, and presentations, it became very clear that a published book was necessary to address the concern about climate change and sustainability of the church and the academy

I thank the World Council of Churches for their dedication and work for climate justice. The WCC has placed high emphasis on the importance of creation care, social justice, and sustainability. It has shared its deep social concern by responding to the heightened alarm resulting from the increase of pollution and carbon waste and how it affects the people around the globe who are the most vulnerable.

My deepest thanks to Dr Guillermo Kerber, who has coordinated the work on Care for Creation and Climate Justice at the WCC for more than ten years. He understands the tensions and dynamics of how churches work, or resist working, toward climate justice. His deep theological understanding and approach to climate justice helped launch this book project. He provided guidance and support for this project.

I thank Michael West, the editor at WCC Publishing, who showed great interest and understood the necessity of this project right from the start. His constant care for this project from the beginning to its publication is commended and appreciated. His wisdom and knowledge helped strengthen this book and make it a reality.

I express my gratitude to each of the contributors to this book. Many of the writers are part of the WCC Climate Change Working Group and are engaged in their own churches and organizations on working toward climate justice. Some of the contributors are not part of the Climate Change Working Group, and I appreciate their faithful work on climate justice and their contributions to this book. Each writer showed support and patience with the editing process.

Lastly, I thank my family. My sister, my brother-in-law, my nephew Matthew, and my niece Naomi supported me and encouraged me. Naomi was helpful in her editorial suggestions and assistance. My husband, Perry, showed faithful support in helping me start, write, edit, and complete this book. He stood by me as I showed despair and loss of hope during the difficult times of finishing it. I thank my children for their constant wonder and amazement in my writing, research, and work. My daughter, Elisabeth, who is sensitive to environmental issues, is concerned that such a book on sustainability be published. My two boys, Theo and Joshua, accompanied me to various climate-change events held by the WCC. My youngest, Joshua, was with me in Wuppertal, Germany, when we decided to write this book. To my family, I owe my deepest gratitude and love.

This book is dedicated to my children, Theodore, Elisabeth, and Joshua, and to their generation, who are experiencing the effects of climate change and in whose hands we lay our hope for climate justice.

CONTRIBUTORS

Elias Crisostomo Abramides, born in Argentina, is a licentiate in chemical sciences (University of Buenos Aires) with an expertise in environmental studies. As a member of the Ecumenical Patriarchate Delegation he attended the WCC assemblies in Canberra, Australia; Harare, Zimbabwe; Porto Alegre, Brazil; and the International Ecumenical Peace Convocation in Kingston, Jamaica. As an advisor to the WCC he participated in the WCC Assembly in Busan, Republic of Korea. In Argentina he is Ecumenical Officer of the Greek Orthodox Metropolis of Buenos Aires and South America, and executive secretary of the Ecumenical Commission of Christian Churches in Argentina (CEICA). Within the WCC he has been commissioner of the WCC Unit III on Justice, Peace and Creation. He is a founding member of the WCC Working Group on Climate Change and WCC Designated Contact Point to the UN Framework Convention on Climate Change (UNFCC) Secretariat.

Lucas Andrianos (Luc Aurelien Andriantiatsaholiniaina) received his Ph.D. from the department of production engineering and management at the Technical University of Crete, specializing in environmental sciences and sustainable development. Since 2006, he has worked for the Foundation for Research and Technology Hellas and was head of the Institute of Theology and Ecology at the Orthodox Academy of Crete (OAC). He is the founder and editor of three series of environmental books on "Ecological Theology and Environmental Ethics" (ECOTHEE), "Conservation and Sustainable Use of Wild Plant Diversity" (CSUWPD), and "Sustainable Alternatives for Poverty Reduction and Ecological Justice" (SAPREJ). Since 2011, he has been collaborating with the World Council of Churches as sustainability consultant for the Greed Line study and climate justice within the framework of the "Poverty, Wealth and Ecology" initiative. With the support of the WCC, he organized the second international conference on "Sustainable Alternatives for Poverty Reduction and Ecological Justice" in Madagascar in April 2014, and the fourth "Ecological Theology and Environmental Ethics" (ECOTHEE) conference at the Orthodox Academy of Crete in 2015.

Simon I. Awad is the executive director of the Environmental Education Center (EEC) in Palestine of the Evangelical Lutheran Church in Jordan and the Holy Land, which works in the fields of education, advocacy, protection of biodiversity, care of creation, and climate change. A key element of his work is providing education in an attempt to protect and restore eco-systems in Palestine. The author or co-author of several books, he is also an activist in eco-justice issues, human rights, and voluntary work. In charge of running several local development institutions in Palestine, Mr Awad has represented the EEC in many local and international forums.

Marina Ngursangzeli Behera is professor of ecumenical missiology at the Ecumenical Institute in Bossey, Switzerland (affiliated with the University of Geneva). Dr Behera belongs to the Mizoram Synod of the Presbyterian Church of India. She obtained her M.Th. in missiology from the North India Institute of Post Graduate Theological Studies, a joint programme between Bishop's College, Kolkata, and Serampore College. Her D.Th. in the history of Christianity was earned from the Federated Faculty for Research in Religion and Culture in Kottayam, Kerala, India. Before being appointed to the Ecumenical Institute in 2012, Behera was associate professor and chairperson in the department of the history of Christianity at the United Theological College in Bangalore, India.

Richard Brand has worked since 2012 as an environmental officer for the Evangelical Church of the Rhineland, where he organizes eco-management training courses for congregations and is developing a climate concept for the territorial church. He is an economist whose main fields of interest are international climate policy, environmental economics, and development issues. He has worked for several years in Germany and abroad in the area of development cooperation.

Katja Breyer has worked since 2011 as a development-policy officer for the Evangelical Church of Westphalia. She is a forest engineer and has been working for many years within church institutions on environmental and development issues.

Rev. Klaus Breyer is the director of the Institute for Church and Society. He has been working for many years as environmental officer of the Evangelical Church of Westphalia, implementing many projects and lobby activities on climate protection.

Meehyun Chung is a minister of the Presbyterian Church in the Republic of Korea (PROK). She studied German literature, philosophy, and Protestant theology in Seoul. In 1993 she received her Ph.D. in Basel, her thesis being on Karl Barth, Josef Lukl Hromadka, and Korea. In 2006 she was awarded the Karl Barth Prize of the Union of Protestant Churches within the EKD for her doctoral thesis and other articles. She received the Marga Bührig Award in 2013. She served as vice president for the Ecumenical Association of Third World Theologians (EATWOT) and worked as the head of the Women and Gender Desk at mission 21, Protestant Mission Basel, Switzerland, between 2005 and early 2013. Since 2013 she has been teaching as associate professor at the United Graduate School of Theology of Yonsei University, Seoul, Korea.

Ernst Conradie is senior professor in the department of religion and theology at the University of the Western Cape in South Africa, where he teaches systematic theology and ethics. His work is mainly in Christian eco-theology and ecumenical theology. His most recent monographs include *Saving the Earth? The Legacy of Reformed Views on "Re-creation"* (LIT Verlag, 2013) and *The Earth in God's Economy: Creation, Salvation and Consummation in Ecological Perspective* (LIT Verlag, 2015). He is a lead editor (with Sigurd Bergmann, Celia Deane-Drummond, and Denis Edwards) of *Christian Faith and the Earth: Current Paths and Emerging Horizons in Ecotheology* (Bloomsbury/T&T Clark, 2014).

Lucy D'Souza-Krone was born in Goa, India, rooted in the Catholic Church, worked as a teacher, and later engaged in rural education and development for women and children. She learned painting from Jyoti Sahi and lived in the Art Ashram INSCAPE near Bangalore. In 1989 she painted for Misereor in Germany the Lenten Veil *Biblical Women: Guides to the Kingdom of God,* which was printed in large quantities and has been seen in many countries in the world. She is married to Andreas Krone, a pastor in Germany. She continues to paint and has exhibitions and workshops in Germany, other parts of Europe, and beyond. Her recent series, "Art for Climate Justice," aims to raise awareness of the challenge of climate change.

Rev. Henrik Grape is an officer on sustainable development in the Church of Sweden. He is a member of the enabling team of ECEN (European Christian Environmental Work) and has attended most of the UNFCCC COPs with the

WCC climate-change group since 2006. He prepared the Uppsala Interfaith Summit in 2008 and the manifesto that emerged from that conference. Most recently, he has been involved in the writing of the Swedish bishops' letter on climate change.

Rev. Fletcher Harper is executive director of GreenFaith, a leading interfaith environmental organization. An Episcopal priest based in the United States, he oversees GreenFaith's Divest & Reinvest Now! campaign, which supports religious divest–reinvest campaigns globally by providing educational and thought-leadership resources, strategic support, and networking. GreenFaith offers a range of programs to help diverse faith communities put their beliefs into action for the earth.

Joy Kennedy is a consultant and long-time activist who has worked for ecological justice through faith-based organizations and NGOs. She is a member of the WCC Climate Change Working Group and the Poverty, Wealth and Ecology Reference Group. She has held eco-justice portfolios with the Anglican Church of Canada; KAIROS, Canadian Ecumenical Justice Initiatives; the United Church of Canada; and the Canadian Council of Churches. She has also served on many boards, such as the Climate Action Network-Canada; Green Communities Canada; the Canadian Business Ethics Research Network (CBERN); Transparency International; and the Canadian Interfaith Conversation. She is a frequent public speaker and has addressed the United Nations several times on climate change, sustainability, and social and economic justice issues.

Guillermo Kerber presently coordinates the work on Care for Creation and Climate Justice at the World Council of Churches, based in Geneva, Switzerland. He holds academic degrees in philosophy and theology (ITUMS–PUG, Montevideo, Uruguay) and a doctorate in sciences of religion (UMESP, Sao Paulo, Brazil). In Uruguay, his home country, he was professor of social ethics both at the National and Catholic Universities in Montevideo and the director of an NGO working on development and human rights. He is author of various books and articles on climate change, human rights, and international affairs from an ethical and theological perspective, published in Africa, Asia, Europe, and North and South America. He currently teaches at the Atelier Œcuménique de Théologie in Geneva (http://www.aotge.ch) and sits on the board of the Global Call for Climate Action (http://www.tcktcktck.org), Oeku–Church and Environment, Switzerland (http://www.oeku.ch), and the International Network for Human Rights (http://www.ridh.org).

Grace Ji-Sun Kim received her M.Div. from Knox College (University of Toronto) and her Ph.D. from the University of Toronto. She is a visiting researcher at Georgetown University. Kim is the author of nine books, most recently *Embracing the Other: The Transformative Spirit of Love* (Eerdmans, 2015). She is a co-editor with Dr Joseph Cheah for the Palgrave Macmillan book series "Asian Christianity in Diaspora." She serves on the American Academy of Religion's (AAR) Research Grants Jury Committee; is a co-chair of the steering committee for AAR's Women of Color Scholarship, Teaching, and Activism Group; and is a steering-committee member of AAR's Comparative Theology Group and the Religion and Migration Group. She sits on the editorial board for the *Journal for Religion and Popular Culture* and is a referee for two other journals: *Journal of Race, Ethnicity, and Religion* and *The Global Studies Journal.* She is an ordained minister of word and sacrament within the PC(USA).

Jochen Motte was born in Germany and has studied Protestant theology in Tübingen, Bern, and Wuppertal. In Wuppertal he worked at the theological seminary, where he wrote his thesis on biblical theology. An ordained pastor of the Evangelical Church in the Rhineland, since 1993 he has worked as executive secretary in the United Evangelical Mission Communion of Churches in Three Continents, where he presently heads the department for Justice, Peace and the Integrity of Creation.

Jea Sophia Oh is assistant professor of philosophy at West Chester University of Pennsylvania. She has developed comparative postcolonial eco-theology in the area of constructive theology to combine her research areas of comparative theology, process theology, environmental ethics, postcolonialism, and feminism. Her book *A Postcolonial Theology of Life: Planetarity East and West* (Sopher Press, 2011) is the first approach to bridge postcolonialism and ecological theology with the use of Asian spirituality as the philosophical underpinning for the argument that all forms of life are sacred.

Pawel Pustelnik is a Ph.D. student at Cardiff University, UK. He is researching issues related to global governance and climate change, with a special focus on emissions-trading systems. For the past five years he has been engaged in the youth ecumenical movement through the Ecumenical Youth Council in Europe and the World Student Christian Federation–Europe. He is also actively working with Jewish organizations on matters connected to interfaith dialogue and Jewish heritage.

Eva-Maria Reinwald is consultant for climate justice at the Evangelical Church of Westphalia. She is a political scientist and focuses on educational work. She is a contact person for the Platform for Climate and Development.

Raphael Sartorius, a theology student, is collaborating with Bread for the World—Protestant Development Service (Germany).

Adrian Shaw has been the climate-change officer for the Church of Scotland since 2007. A geographer by education, he worked for many years in local government and has been closely involved with environment NGOs, including Friends of the Earth Scotland and WWF Scotland. He is secretary to the charity Eco-Congregation Scotland and a member of the Climate Change Working Group of the World Council of Churches.

Siddhartha is a writer and social activist and is the founder of the interreligious centre, Fireflies Ashram, an earth-spirituality ashram, based in a village outside Bangalore, India. He is also the coordinator of the Meeting Rivers programme, a global platform of religious, spiritual, and secular actors who bring fresh understandings, experiences, and solutions to the interrelated challenges of personal change, social transformation, and ecological engagement.

Ilkka Sipiläinen works as secretary for Social Responsibility in the national office of the Church Council of the Evangelical Lutheran Church in Finland. He is a pastor who has studied theology and social sciences. His hobbies include cycling, sailing, and eco-theology. Sipiläinen was nominated for Finnish Pastor of the Year in 2011.

Larisa Skuratovskaya, M.D., Ph.D., is a scientific secretary at the Institute of General Pathology and Pathophysiology of the Russian Academy of the

Medical Sciences in Moscow, Russia, a professional physician, and also a graduate of the human-rights department at Columbia University, New York. She is involved in advancing gender aspects of human rights and protection of the environment within the framework of several international organizations such as the International Women's Forum, International Museum of Women, International Physicians for the Prevention of Nuclear War, and the WCC Climate Change Working Group. She holds as a priority the promotion of education and tolerance. The author or co-author of more than 50 articles and six books, she is a representative of Russia for the Young Global Leadership Foundation.

Dietrich Werner is senior advisor for theology, ecumenical education, and research at Bread for the World—Protestant Development Service (Germany).

Nino Zhvania, M.D., Ph.D., is an associate professor at Tbilisi Petre Shotadze Medical Academy, president of the Georgian Medical Women's Association (http://www.gmwa.org.ge), and editor-in-chief of the international scientific journal *Actual Topics on Women's Health* (http://www.actualtopicswomenshealth.org).

Part One

Churches Respond to Climate Change

CHAPTER 1

Doing Theology in a New Way

A Swedish Perspective on the Climate-Change Challenge to Churches

Henrik Grape

The Churches' Specific Roles and Contributions in Climate-Change Work

To be church is to be contextual. This understanding sums up why we as faith communities must act on climate change.

The context today is that we are living in a world where humanity is in a new position in relation to planet earth. For the first time humans are behind the steering wheel of the whole planet and its eco-system's future. We not only have our own destiny in our hands but also that of its future generations. Over the last century the impact from human activity has gone from having marginal effects on the planet, to having more serious regional effects on the environment, until today, when we have reached a moment where we as humans are threatening the planet's whole eco-system. We are close to or, in some respects, over the boundaries of what the planet can stand without risking the future of its life-supporting eco-systems. The risks of irreversible changes are higher than ever and we as humans are the ones who have the future in our hands. Since the situation that humanity has created places us in the planet's driving seat, we have to decide how we are to navigate for its future. This has never been the case before. One of the most urgent threats to the planet, with consequences for its various eco-systems, is climate change. The window for acting on climate change is shrinking very fast and we must act now if we want to have a chance of alleviating the effects of an already-changing climate.

This must be reflected in our ways of being church or faith communities. To be church is to reflect the gospel in our time. Out of what we know and what we see, we are driven to act. When we realize that those most vulnerable to climate change are the poor, we must speak out. When we know that the effects on climate change will have a great impact on the children of today and future generations, we should recognize that the decisions taken today must safeguard their future. When we hear the cry from the creation, we must change our ways and start to walk another path. From what we know and from what we witness, we hear the gospel of liberation and reconciliation that must guide us as faith communities. These understandings must be linked to visible action and commitment to change. Therefore, the Church of Sweden is engaging in the work of environmental issues.

The Environmental Work in the Church of Sweden

The UN Earth Summit in Rio de Janeiro 1992 was a milestone in global environmental work. Even if the first meeting on environment took place in Stockholm, Sweden, in 1972, the real step forward was the Rio meeting in 1992. It occurred at a time when the world was in a state of "possibilities." It seemed possible to change when the Cold War had come to an end. Many important decisions were made at the summit on biodiversity, the precautionary principle, Agenda 21,[1] and the United Nations Framework Convention on Climate Change (UNFCCC). Even if the follow-up was broken after the turn of the millennium by the war on terrorism and, later, the worldwide financial crisis, the decisions are still relevant and still give us something to hang on to when it comes to taking steps forward on climate policies and issues of sustainability.

The WCC message from the 1992 Rio summit met with a response from people engaged in the environment inside the Church of Sweden. In 1992, volunteers who saw the important connection between being church and taking responsibility for the creation started a network, led by the bishop of Linköping, Martin Lönnebo, and biologist and author Stefan Edman. Ten years later, the network was adopted as a work that should be done within the structures of the Church of Sweden.

From this network inside the Church of Sweden, different tools began to develop. The need for systematic eco-management was early indentified as an important way to change the structures of Church of Sweden to take more responsibility for its own ecological impact. Even though there are systems like ISO 14001 or EMAS (Eco Management and Audit Scheme[2]) to assist in this work, the church also needed to link the question of why we, the church, do this kind of work to the need to root that environmental work into our identity of being church. Coming to this understanding was more important than the ongoing problem that ISO and EMAS require of small parishes more paperwork and a greater amount of money.

With the pattern from ISO and EMAS we constructed a standard for parishes who wanted to earn from the Church of Sweden an Environmental Diploma for Sustainable Development. The standard is structured like ISO standards and requires the congregation to set targets in seven different areas of its administration and work. This includes one area called "pastoral activities," which involves worship services, confirmation groups, baptism, and the like, and the ways in which environmental/sustainable development is highlighted during this part of the congregation's work.

Another thing which distinguishes this Environmental Diploma of Church of Sweden is that all the staff must be involved in a study programme on the environment and churches. This is also the case for all the members of the board of the congregation. This aims to motivate people in the congregations to think and reflect on why we as church are doing this work. When this work started there were checklists of questions to answer as a part of the standard. This was changed and now every congregation has to create its own description of the context and, from that, also prioritize the most important areas in which to set targets.

The environmental diploma is not the easiest thing to cope with. On the one hand, it takes some time for some to understand the meaning of documentation and having a systematic work. On the other hand, this way of working is less sensitive to changes of staff. Changes in routines, for example, will not step backwards due to the documentation and decisions made. The environmental diploma is an ongoing process. It is a systematic environmental work that sets targets; allows for follow-up, evaluation, and communication; and then goes back to setting new targets.

It is not as costly as other systems on the market, but most important is that it involves everything that the congregation works on, not only energy consumption and use of office materials or chemicals. The heart of the work lies in reflecting on the question of why we as a faith community do any particular activity. When this was introduced at the beginning of year 2000 it was not an immediate success, but as time goes by more and more congregations learn the method and see the importance of taking responsibility when it comes to environment—and then it is good to have this kind of tool to offer.[3]

This is occurring at the ground level and it is needed if the church is to be credible in our efforts to act on environmental issues. At the same time, it is important to link all our environmental work to a global view. The environmental work is a part of being one earth and one humanity. What we do locally has impact on the global development. As churches and faith communities, we are sharing this earth. When we are celebrating mass we share the body of Christ. The earth is a part of the creation and when we share bread and wine, we are actually sharing the world. Bread and wine are fruits of the earth and of human work. The eucharist is a symbol of one earth and that we all are a part of this world and the community of life.

Therefore, we must take part in society's decisions. Decisions we make today create the future. We as congregations must take responsibility to lessen our ecological footprint, but there is also an imperative for us to get involved in the world of decision making, nationally, at the European level, and in the global context. Nationally, we have during the last decade, in different ways, tried to reach the decision-makers in the parliament to act courageously on climate change. It is also important to do this on an ecumenical basis and act together as faith communities. As a part of this work, the Christian Council of Sweden has a working group on climate change/sustainable development, which has published a small book with common stances on this subject, entitled *Protect the Earth! That God Loves.* From the different churches there is a common message that we have to act, and to do so out of our identity as Christians. With a document like this it also became obvious that we both have to bring the questions on climate and sustainability into our churches and congregations and also to approach the different parties and actors on the political scene.

Before the UN climate-change summit in Copenhagen in 2009, Archbishop Anders Wejryd, together with the secretary general of Diakonia,[4] approached the leadership of all the political parties, urging them to act more bravely on climate change. The same thing happened in 2014 prior to the general election for the Swedish parliament, stressing the importance of acting on climate change. Sometimes it is not quite clear why churches are involved in climate-change issues. This has to do mainly with the fact that many still view climate-change issues as only an environmental question. But when it is stressed that this has to do with justice, with lifestyle and ethics, it sometimes changes the conversations and brings forth another important angle on climate change.

At times, climate-change issues are met with the argument from members of a national government inside the European Union (EU) that they as a national government cannot do as much as they would due to their membership in the EU. And, to some extent, that is of course true. Thus, the European Christian Environmental Network (ECEN) has over the years contributed in different ways by meeting with EU officials for the purpose of widening their perspectives on climate change. As voices from faith communities we raise another perspective that involves and includes the perspective of the poor and the need for a transformation to a just and sustainable world. My impression is that our voices are listened to mainly because we are not representing some specific special interests, as many other lobbyists are. The option for the poor should be regarded as a special interest.

ECEN and ACT Europe/Aprodev[5] are the natural venues through which the Church of Sweden can act in the European context. It is important to stress the ecumenical basis for action together as churches. That's why we also often are cooperating with Roman Catholic actors, such as Caritas and COMECE, when we are meeting with EU institutions.[6]

Interreligious Action on Climate Change

The Swedish Archbishop Anders Wejryd decided in 2007 to invite leaders from different world religions to a summit on climate change. Climate-change science did take a step forward during 2006 and 2007 due to the Intergovernmental Panel on Climate Change (IPCC) fourth assessment report, the Stern Review on the economics of climate change, and the popular movie on the topic, *An Inconvenient Truth*, featuring Nobel Peace Prize laureate Al Gore. The huge impact that rising temperatures could have on the earth and especially the most vulnerable was clearer than ever, and so, in response, the archbishop invited around 30 leaders from different faith traditions that also have engaged in climate and sustainability issues. These leaders signed the Uppsala Manifesto, which expressed concern over the inaction and the necessity to have a new treaty at the 2009 Copenhagen summit to replace the earlier Kyoto protocol. At the same time, these leaders also commited themselves to take action in their different religious and cultural contexts, to challenge political and business leaders, and to do so out of a deeply spiritual conviction that this is about acting for justice, peace, and hope for the future, in love and solidarity with all human beings and the whole of the creation.[7]

The Uppsala Interfaith Climate Summit took place in November 2008 just before the UN Conference of the Parties (COP 14), and a booklet containing the manifesto was distributed to all delegations when Archbishop Wejryd delivered the WCC message to the plenary of the UNFCCC. The Uppsala summit was important for all who attended because of the multitude of seminars and other events during its two days. The added value of interfaith engagement was clearly expressed. During the ceremony of signing the manifesto there where blessings from all world religions, and children brought a green globe to the center of the Uppsala Cathedral as a symbol of the earth and the coming generations that will inherit it.

There are always reasons to be critical of summits like this and pompous statements that oblige no one to anything. The Uppsala summit could have been just another meaningless meeting. But even if the media coverage was not that high, there was still a commitment among those who signed it, and the document has been used in different contexts and referred to in other documents. Signatories on the manifesto are often eager to take part in climate-change activities. As examples of follow-up on the summit, the archbishop took part in the 2009 Asian Muslim Action Network interfaith meeting, at the invitation of its general secretary, Mohammed Abdus Sabur. The manifesto was translated into Arabic, and the Environment Educational Center of the Evangelical Lutheran Church in Jordan and the Holy Land hosted a seminar on climate change in 2010 in Bethlehem. The Uppsala Manifesto was also referred to in the statement "Towards Rio + 20 and Beyond—A Turning Point in Earth History," developed for the Rio + 20 meeting in 2012. In 2013 there was a joint meeting in Washington between the primates of Church of Sweden, the

Episcopal Church in USA, and the Evangelical Lutheran Church in America on climate change, resulting in a joint statement. Each of these meetings might not be impressive in itself, but regarded together and viewed from many faith initiatives on climate change, they create a firm ground for action and speaking out with credibility.

The Uppsala Interfaith Summit was a clear step from the Church of Sweden to be more engaged in the climate-change debate both in action at the local level, on domestic policy and advocacy work, and in international ecumenical cooperation.

The Bishops' Letter on Climate Change

The Church of Swedens' engagement in climate change is the latest step to formulate a joint letter from the bishops on climate change. There is a tradition that the bishops together formulate a letter on issues that are relevant to the vocation to be church. The letter on climate change was published in May 2014. It contains a description of the scientific background, a short history of words on the human relationship to nature, a chapter on theological interpretations of climate change, and some challenges to both church and society.[8]

The increased work on climate change by the Church of Sweden has deepened the theological reflection on being church in a climate-constrained world. It has opened up new ways of joint action with other actors in society and in ecumenical work, locally, nationally, and internationally. Of course, it has not happened without friction. At first, the climate issues were very much regarded as only an environmental issue and the response was mostly to look for technical actions. Now it is clearer that climate change is about transformation of the whole global society. Transformation requires visions and values. One important part is to stress the relation of climate and justice. The boundaries of the earth are set and we know the limits. We also know that some consume a larger part of the resources than others. When we as faith communities come to this discussion, we have taken an important step to act for the vulnerable of the earth.

This discussion triggers a new debate in my context. There are climate skeptics and deniers who will never accept any action against climate change. There are also those within the faith communities as well as the secular society who question why churches are so involved in the climate-change debate.

Some believe that theology should not be involved with politics. This discussion does not only concern climate change. But it is a step forward when climate change is not just seen as a narrow environmental issue. The discussion about in what ways we are active on the political scene must always be there, to look out for pitfalls, but when it comes to climate it is obvious that we as faith communities must act. By way of illustration, let me paraphrase something that former archbishop K. G. Hammar used to say: in every room where important decisions are being made there should be three empty chairs—one for the poor of the world; one for the children, the coming generations; and one for the creation, nature itself. These three categories are enough argument for putting climate justice in the forefront when it comes to taking part in the transformation toward a more just and more sustainable future.

CHAPTER 2

Health as a Human Right

Medical Professionals and Climate Change

Larisa Skuratovskaya, Nino Zhvania,
and L. Cornelius

"Human Rights are what make us human.
They are the principles by which we create
the sacred home for human dignity"
—Kofi Annan[1]

"Health is a precondition, outcome,
and indicator of a sustainable society, and
should be adopted as universal value and a
shared social and political objective for all"
—The Lancet/Oslo Commission[2]

In 2000 the UN Committee on Economic Social and Cultural Rights adopted General Comment 14, on "the right to the highest attainable standard of health," known briefly as "the right to health." In order to increase the value of this right, the UN Commission on Human Rights established in 2002 a Special Reporter on the Right to Health.[3] Its mandate was endorsed and extended by the UN Human Rights Council in 2007. Since that time medical professionals have significantly increased their attention to human-rights education. Medical professionals play a key role in civil-society education, as they are in contact with large numbers of people all over the world.

This essay[4] will offer some definitions on health, human rights, and the role of health professionals in the implementation of the Kyoto Protocol and in the preparations of a new binding agreement at the UN Climate Change Conference of the Parties (COP) in Paris in 2015.

Atmosphere levels of CO_2eq (CO_2 equivalent) have increased dramatically since the industrial revolution, creating the "greenhouse effect" on the earth, in which more heat is stored in the atmosphere. While natural forces can explain some of the changes in atmospheric CO_2eq, science agrees that most of the increases—especially with regard to CO_2—are caused by human activity,[5] especially as a consequence of the use of fossil fuels and other greenhouse-gas producers. The increase of quantities of carbon dioxide in the lower atmosphere has resulted in global warming. Over the past 100 years, the concentrations of carbon dioxide, mainly greenhouse gases, rose by 30 percent in the atmosphere.[6]

Climate-Change Affects Health

Climate change has an important impact on human health and considerably worsens it. Concerning this topic, the World Health Organization (WHO) writes, "over the last 25 years, the rate of global warming has accelerated, at over 0.18°C per decade. Sea levels are rising, glaciers are melting and precipitation patterns are changing. Extreme weather events are becoming more intense and frequent." Rising temperatures have led to heat waves, unstable weather patterns, more frequent violent weather events such as floods, droughts, and storms, higher risks of infectious diseases, and threats to food supplies.[7]

Climate change affects different aspects of human life and especially human health. A few scientists, including Russian scientist Michael Budyko, were among the first to publish their research about influences of human's activities on climate change at the beginning of 1970s.[8] However, it was not until 1992 that climate change became an important topic for the international community and the United Nations Framework Convention on Climate Changes (UNFCCC) was adopted in Rio (1992) and later at the UN Climate Changes

Conference on the Parties in Kyoto (1997). The "climate-change issue" was a "scientific uncertainty" initially, and only in the last decades, because of the adoption of the precautionary principle, have we seen the transition from the perception of climate change as an "uncertain phenomenon" to the present comprehension of " the risks accompanying climate change."[9]

Climate change affects the determinants of health, such as safe drinking water, air, food, and shelter. Special vulnerable groups are children, elderly people, the chronically ill, people living in the small islands, developing states, states with weak health infrastructure, megacities, and mountainous and polar regions.[10] Children living in poor countries are among the most vulnerable to the health risks and health consequences of climate change.

The different manifestations of climate change have various health implications: extreme heat results in a rise in deaths due to cardiovascular and respiratory diseases, especially in the elderly. Stroke and dehydration caused by heat waves are the most common among the weather-related deaths.[11] Air pollution has become an additional cause of asthma, which, according to WHO data, affects around 300 million people worldwide.[12] Scientists' prognosis is that the temperature warming resulting from climate change will increase the unhealthy levels of ground-level ozone and harmful air pollutants, a key component in smog.[13] Ground-level ozone can damage lung tissue and reduce lung functions, and aggravate asthma or other lung diseases. At special risk are children, older adults, outdoor workers, and people with chronic lung diseases.[14] Climate change may affect allergies, prompting the spring pollen season to begin earlier and continue longer. Climate change further enhances the spread of diseases transmitted through food, water, and animals (e.g., deer, birds, mice, and insects) and increases salmonella and other gastrointestinal diseases caused by bacterial growth due to the warmer temperatures.[15]

Frequent rains can cause an increase in waterborne parasites, which can get into drinking water and lead to different gastrointestinal diseases. Water-related infections have seasonal cycles. The peaks of water contamination are associated with the rise of Campylobacter during times of rainfall.[16] The increase in the number of days with high temperatures is associated with insect-vectored diseases. Mosquitoes spread diseases such as West Nile virus, a new strain of which was detected in 2002. High temperatures promote the spread of this virus, which causes serious diseases that can have life-altering results. Likewise, Lyme disease, characterized by headache, fever, fatigue, and skin rash, is prevalent in areas with high temperatures.[17] It's clear that dissemination of water-, food-, and insect-related diseases depends not only on geographical areas and seasons, but especially on a country's infrastructure, health-care system, programmes of monitoring water, and so forth. Countries with a weak health-care system and a weak infrastructure are at special risk.[18]

As already stated, the impact of climate change on food is also very important. Global warming causes the reduction of yields as well as reduced productivity of crops and livestock. Some regions may require irrigation, which is undoubtedly associated with higher costs. Rising precipitation events may delay planting and harvesting.

Global warming results in more plant pests and insects. These pests can also move to other countries, where the temperature may be even warmer. Warmer waters increase the risk of infectious diseases and can reduce the abundance of fish and seafood, which can be of concern to the commercial fisheries.[19]

The Role of Medical Professionals in Preventing Possible Results of Climate Change

Medical professionals—doctors, scientists, and other health professionals—began collecting evidence on the influence of weather on health at the local level in the 19th century, but the connections with global climate changes became noticeable in the 1970s and its impact on human health has considerably worsened since. Comparing the situation after two big storms in the USA five years apart—Hurricane Katrina and Hurricane Sandy—Gary Cohen writes that "the healthcare sector understanding and its ability to respond to climate change is still in a primitive stage."

It is necessary that hospitals be prepared in their operations and facilities in order to play their role during extreme weather events instead of being one of the victims. New Spaulding Rehabilitation Hospital in Boston, which opened in 2013, is one of a few hospitals equipped with on-site power generation, openable rather than fixed windows, and mechanical/electrical equipment on the facility roof so as to be prepared for the impact of extreme weather events.

The next step for medical professionals is to reduce the use of fossil fuels. Health-care systems represent a big part of the economy in developed countries, so reducing fossil fuels through conservation efforts and the use of renewable energies will help hospitals to operate for longer times during and after extreme weather events.[20]

The role of medical professionals is not only to study all possible impacts of climate change on human health and take those into account in their medical practice, but also to educate the various population groups and to raise their awareness on this matter. People's education concerning climate-change impact on health is important, such as education concerning smoking cessation, the relevance of physical activity, viral hepatitis, HIV prevention, and the like.[21] Despite frequent discussions on this topic, the level of understanding of all implications is low. For effective health communication it's necessary to inform people of all potential effects of climate change on their health and about the activities they can undertake to reduce risks. The different population groups need different kinds of communication and information. It's necessary to take into account the level of problems in understanding, cultural and ethnic differences, vulnerability to the health effects of climate change, and specific problems in the area.[22] Not only is education important, but it is also vital to put the research results into practice. A good example is the use of a personal CO_2 emissions calculator that provides tips for use at home, in the office, on the road, and at school. Usage of such tools helps to minimize the harmful results of climate change and to change our daily behaviours.[23]

Gender Implications of Climate Change

Looking toward a new climate agreement ("Beyond Kyoto"), it is necessary take into account the gender aspect of the impact of climate change on human life. Climate change has unequal impact on different groups. "It's increasingly evident that women are at the centre of the climate change challenge. Women are disproportionately affected by climate change impacts, such as droughts, floods and other extreme weather events but they also have a critical role in combating climate change."[24]

Women in poor countries have limited access to health care and education, restricted rights, and a limited role in decision-making processes, and so forth. Women, however, are often primary householders, and they can play an important role in mitigating and adapting to climate change.[25]

WHO provides the following definition of health: "Health is a state of complete physical, mental and social well being and not merely the absence of disease and infirmity." As we can see, the definition is quite narrow and doesn't include the word *economic*. With this definition, it is difficult to understand the gender aspects of climate change and to take action.[26] In the meantime, women's experiences with climate change, adaptation, and livelihood resilience under multiple stressors have been largely disregarded within the broader climate-change literature, despite the growing attention on gendered knowledge systems, skills, resources, and specific coping, mitigation, and adaptation strategies that women have employed in the face of climate influence. Gender-based inequalities make women and men vulnerable to climate change in different ways.[27]There are, however, a few studies, for example, about gender differences in environment behaviors, which we will briefly describe below.

China represents the third-largest economy and the highest level of national carbon dioxide emissions compared to other nations across the globe. Yet, little social-science research has focused on the environmentally oriented behaviors of the Chinese, which is key to understanding the levels of environmental impact. Researchers have identified a pattern of gender differences similar to common findings in the West. Women demonstrated greater participation in environmental behaviours at home (e.g., recycling), while outside the home (e.g., environmental organizations) no gendered patterns were exhibited. At the same time, Chinese women expressed lower levels of concern than men, a finding opposite most Western studies. Also, distinct from other settings, in China a higher level of knowledge regarding environmental issues did, indeed, translate into pro-environmental behaviours—thereby not exhibiting the knowledge-behavior gap demonstrated elsewhere.[28]

In the USA, a study tested theoretical arguments about gender differences in scientific knowledge and environmental concern using eight years of Gallup Organization data on climate-change knowledge and concern in the U.S. general public. Contrary to expectations of scientific-literacy research, women convey greater assessed scientific knowledge of climate change than men. Consistent with much existing sociological scientific research, women underestimate their climate-change knowledge compared to men do. Also, women express slightly greater concern about climate change men, and this gender divide does not account for gender differences in key values and beliefs or in the different social roles that men and women perform. Modest yet enduring gender differences on climate-change knowledge and concern within the U.S. general public are an intriguing subject for research.[29]

A gender study from Mexico shapes our understanding of the mitigation-environment association in both origin and destination communities. Using quantitative and qualitative data, researchers juxtaposed these gender dimensions for labor migrants—comparing Mexico's southern Yucatan origin communities with the destination communities of Vermont (USA). The study illustrates how, in southern Yucatan, circular transnational migration alters pasture, maize, and chili production in a peasant field-forest system. Gender norms condition the land-use decisions of migratory households to keep women out of agricultural fields, but in turn may be modified in unexpected ways. With men's migration, more women assume aspects of land management, including in decision making and supervision of hired farm labor. In comparison, in Vermont a largely male migrant labor force help maintain an idealized, pastoral landscape with gender deeply embedded in how that labor is constructed and managed.[30]

In Arab countries there is a different situation, because a few formal regulations prohibit women from owning land. Islamic law protects women as independent legal persons entitled to own land, property, and money in their own names, regardless of marital status. However, these protections are sometimes weakened by a lack of awareness among rural women, particularly concerning practices of land acquisition and titling, and also concerning their rights, customary discrimination, and the persistence of the cultural idea that land is owned by men. Researchers write that "shortcomings in gender-disaggregated data related between the link between gender and adaptation in Arab countries is a fundamental challenge because it is limited across the region, especially in rural areas and innovative data collection method that capture gender-relevant information is not commonly used as it could be." Women's vital roles in agriculture, as household managers, and as stewards of natural resources and ecosystems are not accounted for in national data collection and statistic tools.[31]

During the years since the session of the Conference of the Parties (COP) 8, held in New Delhi in 2002, and in a more formal way at COP9 in Milano, and most visibly in Doha (2012) and Warsaw (2013), COP has adopted documents that support options and ways to advance the gender balance in climate change policy. For instance, a gender perspective in adaptation strategies means to empower women to identify the specific needs of the most vulnerable and to help them to participate in decision-making processes. Women also should equally participate in the use of financial resources related to climate-change impact. Women as householders can participate in mitigation projects by using clean energy for lighting or cooking. Improving women's livelihoods and health allows them time to commit to other issues.[32]

In 1997, when the third Conference of the Parties (COP3) accepted the Kyoto Protocol, we took part in the international campaign initiated by doctors from the USA and Great Britain, titled "Medical Warning—Global Warming." At the beginning of the meetings in Kyoto we collected more than 100 signatures of doctors from Russia, and more than 5,000 signatures from all countries which have taken part in the campaign. We held two press conferences in Kyoto during the UN Conference of the Parties, and on 6 December 1997 an article in the *New York Times* stressed the consequence of climate change on health. This campaign was modeled on the "International Physicians for the Prevention of Nuclear War" movement, which in the early 1980s played a significant role in the subsequent termination of the Cold War.

Conclusion

For more than 30 years we have had strong evidence that climate change has been developing. Communities have little time to develop a common understanding in order to take responsibility for the problem. The record-breaking summer of 2013 showed us that we have no time to delay our common action for the protection of our earth. We affirm our strong determination to implement the 2015 global agreement—the COP21 Paris Agreement—that is ambitious, inclusive, and reflects global challenges. We welcomed the Climate Summit of the UN Secretary General in September 2014 and his invitation to all parties to prepare for ambitious contributions and to deliver concrete action, both to reduce emissions and to strengthen resilience. In June 2014 the Brussels G7 Summit Declaration was published; climate change is one of the most important parts of this document, containing a few very important goals, including health-related issues.[33]

Management of the health effects of climate change will require inputs from all sectors of government and civil society, collaboration between many academic disciplines, and new ways of international cooperation that have heretofore eluded us. Involvement of local communities in monitoring, discussing, advocating, and assisting with the process of adaptation will be crucial. An integrated and multidisciplinary approach to reduce the adverse health effects of climate change requires three levels of action. First, policies must be adopted to reduce carbon emissions and to increase carbon bio sequestration, and thereby slow down global warming and eventually stabilize temperatures. Second, action should be taken on the events linking climate change to diseases. Third, appropriate public-health systems should be put into place to deal with adverse outcomes. Medical professionals, through their NGOs internationally and locally, must be involved in all three levels of this ambitious approach.

CHAPTER 3

Advocacy and Action in Scotland

The Work of Churches and Eco-Congregation Scotland

Adrian Shaw

This chapter examines the rapid growth of interest in climate change among churches in Scotland. It suggests why churches have become involved and the role of denominations, particularly the Church of Scotland, in supporting these developments. It also describes the pivotal role of Eco-Congregation Scotland, which from its inception in 2001 has grown to be an ecumenical environmental movement of over 300 churches. These developments have taken place in the context of a small country reasserting its national identity within the United Kingdom, and in churches that have played a role in helping to ensure environmental awareness is part of this new national identity.

The Context: Scotland and the Origins of the Carbon Economy

Scotland is a small country with a population just over five million. As part of the United Kingdom it shares the island of Great Britain with Wales (population 3.1 million) and its much more populous southern neighbour England (population over 53 million). Scotland and England were joined into one nation with one parliament by the Act of Union of 1707, but the Act, while creating the United Kingdom of Great Britain, did allow a degree of self-determination for some aspects of Scottish life, including the independence of the Church of Scotland to maintain its Presbyterian organization and Reformed theology. The independence of the Church of Scotland to determine its own organization and worship has been a distinctive feature of Scottish life ever since, and is one of the historical threads that gives Scotland an identity quite separate from England. Political power, however, remained very firmly with the UK Parliament in London until 1997 when a Scottish referendum gave overwhelming support to proposals to devolve powers to a Scottish Parliament. This came into existence in 1999 and has led to some profound changes in Scottish politics and government. A referendum on whether Scotland should become an independent country was held in September 2014. The referendum led to an impassioned debate, and while the result was a vote against independence, it has resulted in a promise from the UK government that further powers will be transferred from the UK Parliament to the Scottish Parliament.

In common with England, Scotland was in the forefront of urban and industrial development over 200 years ago with spectacular growth of cities, particularly Glasgow. Powered by coal, the industrial and urban transformation of lowland Scotland left its mark on landscapes, communities, and ecology. The new global carbon economy was effectively pioneered in England and Scotland, where innovations in coal mining and the application of steam power to industry, the factory system, railways, and marine engineering led to massive economic expansion and an equally remarkable growth in the consumption of coal. The inevitable consequence of coal burning was atmospheric pollution in towns and cities and the growth of carbon emissions to the atmosphere. When these historic carbon emissions are taken into account, Scotland probably has some of the highest per capita emissions of any country.[1]

Scotland has had a carbon-based economy for two centuries. Throughout the 19th century the coal industry was dominant, but in the second half of that century there also emerged a small and innovative onshore oil industry extracting oil from shale rocks of the lowlands between Edinburgh and Glasgow. This domestic industry was relatively small in scale, and was superseded quite quickly by oil imported from abroad, but it was notable in that it played a pioneering role in the early history of the oil industry.[2] In the 20th century the industry in Scotland was rejuvenated by the discovery of oil fields under the North Sea, where the first successful drilling operations took place in the 1960s and from where the first oil was piped ashore in 1975. Overcoming the technical challenges of recovering oil from the North Sea, with its deep water and often foul weather, has given Aberdeen, the onshore capital of the UK North Sea oil industry, a global role in the offshore oil industry, so in the 21st century Scotland remains a carbon economy of some significance.

Scotland and Environmental Consciousness

Alongside the urban lowlands of Scotland, with its history of oil and coal, there is another side of Scotland: the rural Scotland of the southern borders with England, and the highlands and islands. Rural Scotland, particularly the highlands and islands, is a thinly populated area of hills, mountains, glens (valleys), lochs (both freshwater lakes and sea inlets), and rugged islands. Romanticized by the writings of Sir Walter Scott in the early 19th century, the rural landscapes of Scotland have subsequently become a major tourist attraction. With extensive areas of open country, endless opportunities for outdoor activities, and wildlife largely absent from much of the crowded UK, the highlands and islands are often referred to as a wilderness. Curiously, the rural landscapes of the North and West are neither truly natural nor unspoiled. Much tree cover was removed by agriculturalists before the Christian era, and biodiversity has been reduced by millennia of human activity.[3] According to one source, "By the time the Romans arrived, over half of our native forests had been lost."[4] In more recent centuries the human population has also been reduced in many remote areas, with once-thriving rural populations declining due to emigration, including the forced emigration of the "clearances," leaving many remote areas with a lower population now than 200 years ago.[5] In many ways the iconic "wild" landscapes of highland Scotland have been largely shaped by human activity over countless generations.

The contrast between nature and urban life became a passionate concern of perhaps the greatest environmentalist to be born in Scotland. John Muir (1838–1914) was born in Dunbar in East Lothian and grew up in a profoundly religious family. His father was a Campbellite Presbyterian and, in search of a better life, took his family to Wisconsin, USA, in 1849.[6] John Muir reacted to the natural environment of America with enthusiasm and delight, first around his family home in Wisconsin and later in his adult life in his wanderings in the forests and mountains of California, Alaska, and beyond. He wrote about his journeys in influential newspaper articles and books, lobbied politicians and government, and helped form the Sierra Club, one of the earliest environmental lobbying organizations. Repeatedly in his writings he stresses the sanctity of nature; that the mountains and forests are nature's true temples. In his mind, deeply rooted in Scottish Calvinism, exposure to America's mountains and forests helped to shape a new environmental consciousness that he shared with others and which became immensely persuasive. It is striking that the environmental movement took root in North America much more quickly than in Scotland, and it was not until a century after his death that John Muir came to be more widely appreciated in his native land. This reawakening of interest is part of the growth of environmental consciousness taking place in Scotland today.

Environmental Awareness and Action in Scottish Churches Today

The Church of Scotland has been an active participant in the debate on climate change since the 1980s. In 1989 the Society Region and Technology (SRT) project of the church published a small volume of essays on climate change, entitled *With Scorching Heat and Drought? A Report on the Greenhouse Effect.*[7] This publication includes a number of themes: an explanation of the science of the greenhouse effect; the impact of rising levels of carbon dioxide on climate and the likely effect of increasing temperatures on ice caps; the policy options open to government for adaptation and mitigation; and the need for a theological and practical response by churches. It is interesting that over 20 years ago this publication anticipated themes that have become the agenda for action by churches, and while our understanding of the issues has become more refined, the essential parameters remain the same. Some of the main developments in Scotland are summarized briefly in the table below.

Date	Developments in Scotland
1989	Publication of *With Scorching Heat and Drought* by Society, Religion, and Technology Project
2002	First Eco-Congregation Award made in Scotland to Callander Kirk
2006	100th eco-congregation registered in Scotland: Dunnichen, Letham, and Kirkden Church in Angus
2007	Energy for a Changing Climate report to Church of Scotland General Assembly Responding to Climate Change project established by Church of Scotland
2008	200th eco-congregation registered in Scotland: Strathfillan Church l/w Glenorchy and Innishael Church, Argyll
2009	Ecumenical climate-change service led by the moderator of the General Assembly, Cardinal, and Primus of the Scottish Episcopal Church before "The Wave," a march through Glasgow to raise awareness of the Copenhagen COP
2010	Eco-Congregation Scotland established as an ecumenical charity
2013	300th eco-congregation registered in Scotland (Priestfield, Edinburgh) Appointment of Environmental Chaplain to ECS

Table 3.1: Environmental Action by Churches in Scotland: A Brief Chronology

Eco-Congregation Scotland

Pivotal to the development of environmental awareness and action by churches in Scotland has been the ecumenical movement and charity called Eco-Congregation Scotland (ECS). The idea of eco-congregations originated in England in the late 1990s under the umbrella of Churches Together in Britain and Ireland (CTBI) and an environmental charity, EnCams. A minister of the United Reformed Church, Rev. David Pickering, prepared original materials and suggested a structure similar to that for Eco-Schools, which encouraged churches to make a commitment to environmental action and to work toward an award that would recognize and celebrate their achievement. In Scotland the idea was quickly adopted, particularly due to the commitment of Margaret Warnock, a member of the congregation at Callander Kirk, which became the first award-winning eco-congregation in Scotland in 2002. With support from the charity Keep Scotland Beautiful and the Church of Scotland SRT project, other registrations soon followed. The movement grew quickly in England and Scotland, with 100 congregations registered in Scotland by 2006. Toward the end of its first decade it became clear that it needed to be put on firmer foundations, and what had been an informal association became a formally incorporated company with charitable status and a board of directors.

Becoming an Eco-Congregation

To encourage churches to join, the process of becoming an eco-congregation has been simplified to make it as easy as possible. All that the members of a congregation have to do is to fill in a simple registration form, get it approved by the church decision-making body, and post it to the ECS office. The invitation to join is open to all congregations in Scotland and the number of registrations has continued to grow steadily. Many congregations, however, still have reservations about joining, suggesting that some congregations lack confidence to take this step or believe that they are not good enough to become eco-congregations. Overcoming this lack of confidence remains a challenge.

Upon registering, a church is encouraged to carry out a simple environmental check-up, using a checklist they can download from the ECS website.[8] This introduces three areas that churches can explore and allows congregations to establish a benchmark from which they can take action. The three areas are:

Spiritual Living, Practical Living, and Global Living. For each area the ECS website provides resources that are free for churches to access—resources that are designed to assist congregations develop an understanding of God as Creator and offer practical ways in which they respond to this. "Ideas for Action" are offered as a resource to help churches explore issues, learn together, and grow as an eco-congregation.

Practical support is provided through a scheme of mentoring, which is available to any new registrations and links them to a nearby eco-congregation with the experience and confidence to help support churches in their first steps. Congregations are also encouraged to take part in local networks, of which there were 18 across Scotland in 2014. Networks are encouraged to meet two or three times a year to provide a forum to share news, discuss how they are getting on, and to learn about new developments. Since 2008 there has been an annual gathering with keynote speakers and workshops on themes including climate change, food and the environment, and also a range of other regional or thematic events.

The growth in the number of eco-congregations in Scotland includes churches from ten denominations, which reflects the ecumenical nature of the movement and shows how interest in environmental issues is widespread among congregations. This strongly suggests a broad-based rediscovery among Christian churches of what it means to care for creation and how this can become an integral part of spiritual development. It also suggests that the eco-congregation format, which is challenging but supportive, is one that can help promote this journey of rediscovery. This format is not unique to Scotland. Eco-congregations developed first in England and similar movements can be found in churches in Germany, Hungary, the USA, and in churches across the continents.[9] It is a format that can be adopted and adapted in many settings, with results that are often impressive.

Responding to Climate Change

There is, of course, more to caring for creation than climate change, but it is climate change that has emerged as the most important single environmental issue in the Church of Scotland's work. This may be for three reasons. First, as noted above, there has been a longstanding interest in climate change in the work of the SRT project, and the SRT committee has returned to the theme on more than one occasion since 1989. Second, stories from partner churches about the impact of climate change on countries around the world, particularly in the global South, have created pressure for action within the church. For example, reports from the Church of Bangladesh have brought home to congregations how climate change is changing lives. James Pender of the Church of Bangladesh has written of the impact of climate change on the country, including the risks it faces associated with climate change, with a combination of droughts, floods, and cyclones all posing a threat to life and well-being. Rising sea levels pose particular risks, as millions of Bangladeshis live on ground that is less than one meter above sea level.[10] This message was made clear in a presentation to the general assembly of the Church of Scotland in 2006 by Bishop Michael Baroi of the Church of Bangladesh. This presentation, with its accumulation of evidence from other countries, including Malawi, Ghana, and other partners, particularly Christian Aid Scotland, has had a persuasive emotional impact on the church and its members.

A third issue was the growing debate in civic society in Scotland about environmental issues, and climate change in particular. Scotland has a vibrant and vocal voluntary sector with NGOs, including Friends of the Earth Scotland (FoES), the Royal Society for the Protection of Birds (RSPB), and the World Wide Fund for Nature (WWF), that have been campaigning vociferously on climate-change issues for some time. There has been active church involvement in this debate, for example the important role played by Fred Edwards (1931–2008), who was chair of Scottish Environment Link and an active member of the Church of Scotland, and contributions from leading churchmen such as Canon Kenyon Wright of the Scottish Episcopal Church or Cardinal Keith O'Brien, who was a tireless traveller promoting awareness of the global impact of climate change. These links have helped promote awareness in churches of environmental issues and helped convince environment NGOs that the churches have a role to play. There has been little evidence in recent years in Scotland of suspicion between church and environmental campaigners, and few church members or ministers express opposition to linking church life with environmental issues. Rather, the reverse is true, and many congregations have been ready to embrace the idea that caring for creation should be at the heart of Christian life and that both spiritual and practical action are required.

How Can the Church Respond to Climate Change?

In 2007 the Church of Scotland responded to these growing pressures for action by creating the project *Responding to Climate Change*, and appointing a project officer to take forward the work. The project represented the culmination of the pressures noted above as well as a new commitment that was both symbolic and practical. The project offered the church the opportunity to reflect how it could respond to this new global challenge, and how a response to climate change could be rooted in the mission and teachings of the church. Three areas of work were identified at the outset. The first was to help congregations learn about climate change and support them in facing up to the many challenges it poses for them in their worship, in their lives, and in their communities. In developing this work, the partnership with Eco-Congregation Scotland has been of great importance. Through guidance, worship materials, conferences, annual gatherings, and network meetings the structures created by Eco-Congregation Scotland have given ample opportunity to spread messages and promote discussion.

The annual gathering of Eco-Congregation Scotland, for example, has given a platform for speakers, including the Scottish government environment minister Stewart Stevenson, the writer Alastair McIntosh, and the former UK climate diplomat John Ashton, to inspire congregations to action. The Church of Scotland encouraged eco-congregations to adopt the idea of carbon-footprinting church energy use, and now, to gain an award, eco-congregations must calculate the carbon footprint of their church buildings. A simple spreadsheet calculator allows congregation to work out the carbon footprint of the energy used in church buildings from their energy bills or meter readings. Congregations going on to seek a second or third award must demonstrate that they have explored how to reduce this carbon footprint. Pointers to sources of advice on energy management and funding, such as the Energy Savings Trust and the Scottish Government Climate Challenge Fund, have encouraged congregations to commission energy audits of their buildings and take action to manage their energy use more effectively. However, the challenge of heating large and often draughty church buildings in a cool, damp climate are considerable and many churches struggle with rising energy prices and the difficulties of retrofitting buildings that are old, often very historic, and protected by listed building status. Some congregations have risen to this challenge by installing new heating systems, insulating or partitioning spaces where possible, and, in some cases, installing new low-carbon technologies like solar panels or heat pumps. However, reducing fuel bills and carbon footprints in churches remains the most challenging part of the new environmental agenda for many congregations.

A second theme of the work has been to work with the councils and committees of the church to explore what climate change means for different areas of the work of the church and to explore new ways of working with ministers and others. One recent development has been the development of a staff "green group" in the Church of Scotland offices at 121 George Street in Edinburgh. This is a group of staff that have come together from different departments and who have lobbied for better facilities for cyclists, much-improved recycling facilities, and a staff business-travel policy promoting use of public transport rather than car usage. The work of the group was enhanced in 2014 when it took the decision to register as an eco-congregation. This is the first time a group of staff working together in an office has become an eco-congregation and demonstrates how the concept is sufficiently strong and flexible to grow beyond churches alone.

A Commitment Driven by Faith

While much of this work is pragmatic and managerial (there is little spiritual about caring for a heating boiler), it remains rooted in faith and is driven by a commitment to care for creation. It is therefore essential that the response of the church to climate change explores what it means to care for creation, and Creation Time has played an important part in this. Learning from the work of other churches in the European Christian Environment Network and especially from a partnership with the Reformed Church of Hungary, the Church of Scotland adopted the idea of Creation Time at its 2011 general assembly. In partnership with Eco-Congregation Scotland, worship resources were first developed for Creation Time in September 2012 and new materials have been added in subsequent years, resources that are made available on the Church of Scotland *Starters for Sunday* website, the ECS website, and elsewhere.[11] The development of these materials has been led by the ECS environmental chaplain, a post created in 2013 which has made a significant contribution to the awareness of creation theology and worship in Scotland. The post is a gift of the United Reformed Church through its Special Category Ministry scheme and has provided a great boost for congregations exploring care for creation in worship.

The third area of work has been involvement in the public debate about climate change in Scotland. The Church of Scotland became a member of Stop Climate Chaos Scotland (SCCS) in 2008 at a time when the Scottish government had published draft proposals for climate-change legislation. SCCS is a very broad coalition of environment and development NGOS, trade unions, student groups, and others. Other faith-based groups in membership of SCCS include Christian Aid, the Scottish Catholic International Aid Fund (SCIAF), Tearfund, and the Scottish Episcopal Church. The Climate Change Bill was introduced into the Scottish Parliament in 2008 and created discussions among campaigning groups, businesses, and others about why and how Scotland should respond to climate change. SCCS campaigned vigorously on targets set for carbon reduction, and in Parliament these were made more challenging than the Scottish government had originally intended, a major campaigning success for SCCS. The churches and other Christian groups supported this lobbying effort, but the Church of Scotland also brought its own particular perspective on the content of the draft legislation. While many groups argued over the need for stronger carbon-reduction targets and mandatory reporting to Parliament on progress, the church, reflecting the debates and concerns expressed by congregations around Scotland, argued that there was limited public understanding of the pressing need for carbon reductions and limited awareness of how the substantial cuts proposed in the bill of an 80 percent reduction in Scotland's total carbon emissions could be achieved by 2050. In order to help face up to the life-changing challenges involved in creating a low-carbon economy, the church argued that there had to be better public awareness and engagement on climate change issues. To this effect the church lobbied for an amendment to the bill to require the Scottish government to promote a public-engagement strategy to help people across Scotland learn about the issues and get involved. An amendment to this effect was promoted and was accepted by members of the Scottish Parliament. The Climate Change (Scotland) Act 2009 was duly passed as amended and work on behavioural change to promote low-carbon lifestyles has subsequently become a major theme of the implementation of the legislation.

It has been a source of pride to the Scottish Parliament and Stop Climate Chaos Scotland that, when passed, the act was one of the most demanding pieces of climate-change legislation in the world, setting targets that demand that Scotland reduce its carbon emissions by 42 percent from 1990 levels by 2020 and by 80 percent by 2050. Achieving these targets remains a huge challenge and there remains no certainty that they can be achieved; interim targets covering the years from 2010 to 2013 have all been narrowly missed. What is certain is that churches have a continuing role in challenging congregations to think about why this is necessary in terms of faith and asking congregations to consider how they can be leaders in their own communities in responding to climate change.

What Have We Learned?

Why Scotland has been fertile ground for this rapid growth of environmental actions by churches is worth considering. In part, it represents a chance coming together of circumstances. The eco-congregation model, developed in England, took root quickly in Scotland with the support of denominations and government. The support of denominations has been of great benefit to the growth in the number of eco-congregations, and the Church of Scotland, Scottish Episcopal Church, and United Reformed Church have all provided moral and material support, as have Christian Aid and SCIAF. Financial support from these organizations and from the Scottish government has given Eco-Congregation Scotland the resources to promote its activities and do new recruitment. Locally, churches of many denominations have responded well to this opportunity. Assessors in eco-congregation award meetings have noted that, for many eco-congregations, active participation has deepened their faith and given them a renewed sense of purpose and new direction. In an age of secularism and declining church membership, such a revival of faith is clearly important. Whether it can have a long-term impact on church life remains to be seen.

There is no doubt that the devolved Scottish Parliament set up in 2009 and the independence referendum of 2014 stimulated political debate and discussion. All these developments have helped boost self-confidence that a small country can take action on big challenges such as climate change. It is perhaps ironic that a country with a large historic carbon footprint and a deeply rooted carbon economy that continues to embrace the coal, oil, and gas industries should be considering a low-carbon future and, for some, equally surprising that the churches should seek to play a significant role in this transformation.

With the Scottish government the churches are concerned to promote awareness and action on climate change. The Scottish government is doing this to ensure it can meet the statutory carbon-reduction targets, while churches and congregations are motivated by faith and the need to care for creation. So, starting from very different perspectives, congregations and government are coming to share a similar vision and one in which churches can play a role in helping Scotland move toward a low-carbon future.

The obstacles to the journey to a low-carbon economy are difficult to exaggerate and might be compared to the great and arduous pilgrimage John Bunyan described in *Pilgrim's Progress*. Consumerism remains a powerful ideology in Scotland, with shopping and its promise of instant gratification almost a religion in its own right. Governments, including the Scottish government, continue to talk about economic growth as the solution to our problems at the same time as they talk about reducing carbon emissions. The mission of the church, and the gospel-driven demand that we set aside our desire for material wealth as an impediment to real life, takes on a new focus when we consider the disastrous climate consequences of a consumer-driven economy. For churches in Scotland the age of climate change is an opportunity to rediscover in faith and in practical action what it means to care for creation.[12]

CHAPTER 4

Youth for Climate Justice

A Case Study

Pawel Pustelnik

This chapter looks at the issue of ecological justice from a slightly different angle than the others in this book, as it asks about the work on climate and ecological justice that is being done by young Europeans. A recent campaign launched by the Ecumenical Youth Council in Europe (EYCE) lends itself as a case of the youth involvement with the issues that this book aims to tackle and provides tangible examples of actions that were implemented in Europe between 2011 and 2013. First, I introduce the organization and explain EYCE's decision to focus on ecology. I next describe the contributions made over the three years and look at the results of the youth involvement. Finally, I analyze the difficulties and challenges that ecumenical youth face related to ecological justice.

The Ecological-Justice Campaign in the Context of EYCE's Mission

EYCE is a Christian ecumenical umbrella organization that involves national ecumenical youth councils, denominational youth councils or bodies, and international Christian youth organizations from 24 European countries. EYCE was established in 1968 and since its very beginning has been closely connected to the work of the World Council of Churches (WCC). After several moves within Europe, today it is based in Brussels, the political heart of Europe. The organization's motto comes from Galatians 5:13, "Called to Freedom, Justice and Peace," which has always been a pathway to thinking about the initiatives with which EYCE engages.

The council has been working successfully on a basis of three-year campaigns since 2004; the one which spanned from 2011 to 2013 was entitled "EYCE Campaign to Promote Ecological Justice." Before the launch of the campaign there were, however, several attempts to approach the topic of eco-justice or ecology in general. For example, in 2004 EYCE organized a summer school on ecology and sustainability titled "Water, Air and Earth—Celebrating Nature, Improving the Future." During that event, for the first time in the EYCE's history, the thematic focus was placed on the responsibility of young people for creation. The overwhelmingly positive feedback made the council's decision-makers think of elaborating on the issues related to ecological justice in more detail. In 2009 the decision was made to work on that issue in the coming years through the form of a campaign. This campaign was run by a group of volunteers called the Campaign Coordination Team, supported by the EYCE's general secretary and an office intern.

The overall aim of the EYCE Campaign to Promote Ecological Justice, approved by the EYCE's General Meeting (the main decision-making body), was "to commit EYCE and its Member Organisations to act responsibly towards Creation and enable the young people from this network to do so."[1] This reflected the issues related to green awareness among the ecumenical youth, their knowledge of the developments concerning the United Nations Framework Convention on Climate Change (UNFCCC) process, and the links between the ecology and climate debate and religious communities.

Even though Europe may hear little about climate skepticism, green technologies are widespread, and young people are well aware of their influences on the planet, the issue of ecological justice still does not enter the mainstream debate. Often, young people are not familiar with the ecological-justice framework and the questions concerning their wider ecological responsibility. Additionally, EYCE wanted to help its network to link the secular "green thinking" and practice with Christian values and scripture, and in a way in which they could grow in their faith. Furthermore, the EYCE's membership context features very diverse communities: it is enough to think of how differently ecological responsibility is perceived in Portugal, Germany, and Belarus, or how significant are the differences between daily routines in large metropolises and vast rural areas of Finland.

To embrace the differences and provide grounding for the activities, several objectives were set for the campaign:

- to raise awareness of ecological issues in Europe and beyond;
- to study and analyze the developments from the United Nations climate conferences in Kyoto and Copenhagen;
- to explore the relation between ecology, economy, and politics, including reviewing ecology issues as a basis for numerous conflicts;
- to empower the organizations and/or individuals to tackle issues connected to ecological justice;
- to enable the organizations and individuals to lobby for a greener Europe;
- to provide practical advice and tips for ecologically responsible lifestyles;
- to develop a policy paper on ecological justice to be presented at EYCE's General Meeting in 2013, which would entail implementing results of the campaign as an integral part of the running of the council.

The ambitious list of targets of the campaign was split into three thematic focuses. During the first year the participants of the events looked at the theological and political perspectives. The second year explored and addressed ecology and justice in the context of relationship between economy, society and politics. The third year was entirely dedicated to the issue of sustainability and the future involvement of the Council with the ecological justice-related issues.

The Theological and Ethical Work of the Campaign

The ecumenical aspect of the work in which EYCE is engaged appears in the theological background that has been the campaign's foundation. A document that retrospectively looks at the campaign's theological underpinnings is called "Policy Paper of the European Ecumenical Youth for the Respect for the Integrity of Creation."[2] This document underlines the importance of "active respect for the integrity of Creation" under the circumstances of a "modern way of life based on overconsumption and [a] paradigm of unlimited growth . . . pollution, resource depletions, deforestation, species extinction and climate change." In the framework of the campaign several biblical references were particularly strongly emphasized. While thinking about the campaign, EYCE was closely looking at God's creation, which shows mercy to us every day. It is the air we need to breathe and the earth on which we need to walk. We, along with our past, present, and future neighbours, are absolutely dependent on God's merciful gift: the creation we are supposed to govern (Gen. 1:28). On the other side, the discourse of neighbour was very present in the theological thinking of the campaign's team. "Love yourself as you love your neighbours" (Matt. 22:39). In the context of ecological justice, the neighbours are not only the immediate ones who experience environmental changes similarly, but those neighbours on the other side of the planet suffering from effects of the overconsumption from the global North. Furthermore, EYCE considers the future generations as our neighbours, too. The theological reflection that was prevalent while thinking of the responsibility for creation was that of violation of good neighbourhood principles as well as misgoverning of the planet that was offered to humans.

The theological perspective is reinforced by an ethical perspective. Therefore, EYCE decided to look also at the question of human dignity and human rights with regard to the victims who suffer or will suffer the consequences of ecological degradation and climate change.

What Has Actually Been Done?

The ideas that were discussed by those who participated in the EYCE's General Meeting in 2009 needed to be translated into concrete actions: identification of people interested in leading the campaign and also intensive fundraising to provide financing for planned trainings and seminars. Shortly before launching the campaign, EYCE, together with its partner organization World Student Christian Federation–Europe, organized a study session, "Climate Justice Now!," that took place at the European Youth Centre in Strasbourg in September 2010.

At the beginning of the campaign there were more questions than answers: How is an ecologically responsible way of thinking and living linked to wealth and poverty? How does the lack of natural resources caused by climate change trigger military conflicts and violence? What is the role and what can be the contribution of the young European Christians to promoting ecological justice?

Given the fact that green awareness of the member organizations was identified as one of the issues to tackle during the campaign, the team decided to approach it in two ways: first, by addressing the needs of the communities and, second, by raising awareness on the individual level. The implementation of the idea was therefore based on running training courses and seminars on one side and working closely with the member organizations on their daily practices. These two were additionally supplemented by so-called greening of the EYCE office, understood as implementing green standards in its work.

Training Courses

The first training course was planned already for the early stage of the campaign: "Taking Steps towards an Ecologically Responsible Society: Theological and Political Perspectives of Ecological Justice" was organized in Kiev, Ukraine, from 3 to 10 July 2011. By blending Christian context with a highly politicized discourse on climate change, the team preparing the training managed to focus on the ecological-justice issue from a unique perspective related to the role of ecumenical youth in their narratives in their respective countries. The Kiev conference fostered discussions about the "greed line" and allowed the participants to take part in a simulation of UNFCCC negotiations. The programme provided the participants also with knowledge and ideas for use in their practical work through sessions on advocacy, project management, and social marketing.

Drawing on the fact that the Campaign to Promote Ecological Justice was designed to induce changes in behaviours, one of the days of the training course was dedicated to actually living a resource-limited lifestyle. The participants and the team refrained from using electronic devices, excessive water consumption, and learned some practical skills in bread-, candle-, and solar cooker-making workshops.

The second training course took place, again in the eastern part of Europe, in Minsk, Belarus, between 24 June and 1 July 2012, and was entitled "Taking

Steps towards an Ecologically Responsible Society: Links between ecology, economy and politics." This session was spent largely on further inquiry into the relationships between the issues the campaign was looking at. The participants were able to deepen their understanding of the global situation through the exposure of UNFCCC data and by pointing out the main challenges that sustainability is facing. Thanks to the fact that some of the participants of this session had already participated in the previous events of the campaign, the training was a platform of mutual learning and exchange of experiences gathered during the period between the two events. As part of the activity, the participants were invited to clean the green areas surrounding the venue. This made them especially aware that illegal garbage dumping is still an issue in some parts of Europe, but this task also helped to empower them to act in their local contexts in similar initiatives.

Eco-justice at a Grassroots Level: "Be Eco Heroes!"

Another important pillar during the campaign was the EYCE's focus on the organization itself in its implementing "Be Eco Heroes! Making EYCE More Sustainable." As mentioned before, EYCE, through the Campaign to Promote Ecological Justice, also wanted to look critically at its network. In its explanation of the reasons for initiating this project, the Campaign Coordination Team said:

> We, ecumenical youth, truly believe that we should care about the Divine Creation, which was given by God to people to dominate it, not exhaustively, but with respect for Creation itself and for the future generations. In order to reverse the negative trend of disrespect in society nowadays, we can change our personal attitudes, we can also become politically active and advocate for changing related political rules.[3]

These ideas have been translated into practical steps that the network is taking. All member organizations have been encouraged to assess their contribution to an eco-just future in their local communities on various levels: challenging lifestyles, influencing the political agenda, inspiring other non-faith-based organizations to act together and become the agents of change. "Be Eco Heroes" became, therefore, a coherent part of the whole campaign.

Having in mind that often the activities of the headquarters may be too distant from the local activities in the member organizations, special care was given to engage individuals at the grassroots level. Furthermore, a manual for the whole network on how to become more sustainable was produced. Thus, EYCE's publications were used to promote the champions—the "Eco Heroes"—and stimulate the others. Below is a list of the seven steps that the organizations received and were instructed to follow.

Step 1	**Commit your organization to turn green** For the success of the overall project, there needs to be a will to change ourselves. We believe that ecological justice is a very important topic for young people, as well as for Christians in general. Therefore, being an ecumenical/Christian youth organization already implies high expectation to join the initiative. However, participation on the project should be in consultation with the organization's leaders. In case there is not overall content and the organization as a whole is not willing to participate, maybe a smaller group of willing groups (e.g., several local youth groups in different cities, a certain group of willing parishes, or a specific group of volunteers etc.) could participate on the project.
Step 2	**Establishing a green committee** If the will and mandate are there, a sort of green committee should be established in order to coordinate the process. This does not mean that this committee will do all the work. The involvement should be as broad as possible. However, a sort of management and responsibility assigned to a certain person is always helpful.
Step 3	**Self-evaluation** The green committee together with help of others should conduct a sort of inventory of the organization with a special focus on ecological impact of your organisation on the environment. Please evaluate: 1. positive aspects of the impact (ways of compensating the harmful effects, or of positive influence on society) 2. negative aspects of your ecological impact (environmentally harmful activities).

Step 4	**Identify achievable objectives** Based on the inventory, identify opportunities for improving both impacts. Then convert selected opportunities into a concrete set of achievable objectives. In doing so, take into account your resources and capacities (financial, personal, etc.). If you do not have resources or capacities needed for some kind of desirable achievement ask for external help (e.g., for an expert consultation in a partner organization), it is recommended to start with simple and easy-to-do objectives which could ideally be developed within the suggested timeframe. If you succeed this time, you can continue with more ambitious objectives next time.
Step 5	**Implement the objectives** Implement the objectives according to the plan developed in Step 4. If you are facing difficulties you cannot manage alone, ask for help—either ask EYCE and its members or a relevant partner in your country.
Step 6	**Evaluate objectives** Come back to your initial inventory and try to evaluate how you have managed to improve the environmental impact of your action. For your future work it might be useful to also evaluate the way it was done. What was done effectively, what could be improved for the next time?
Step 7	In the light of evaluating achievements done so far, elaborate an ecological inventory once again according to the Step 3 and continue from that point on!

Table 4.1 Be Eco-Heroes![4]

The process was designed to be inclusive and to take into consideration the specificity of the member organizations. The method was accompanied with the spiritual part, as underlined in the manual: "Any big change and far reaching decisions need to be rooted in values that you truly believe in. . . . it is important to be sure that the project is consistent with our Christian values and Scripture."[5] Therefore, member organizations were encouraged to promote ecology and sustainability through worship or special spiritual gatherings focused on care for creation, organizing outdoor prayers and services, providing communities with Bible studies on eco-theology, encouraging communities to fast for spiritual as well as ecological reasons, and developing their own prayer/Bible-study material on ecology or sustainability.

Several tangible results came from the programme. The Icelandic member organization Æskulýðssamband Kirkjunnar í Reykjavíkurprófastsdæmum (ÆSKR) decided that in 2013–2014 they would focus on promoting environmental awareness and responsibility in their own context by launching an initiative called "Living the Green Life in God's Garden." This was understood as a programme that would aim to "inform about green values and inspire responsibility toward the environment and God's creation mostly using non-formal education methods."[6] Another example came from Romania, where the Evangelical Lutheran Church in Sibiu became the second organization in the country to receive Eco Management and Audit Scheme Certification, a recognition toward greening the congregation. On a community level, the church decided to launch a series of eco-markets to promote locally produced goods.

Dissemination and Campaign Materials

The Campaign Coordination Team was aware from the very beginning that even if people were interested in the topic that the council was addressing, some of them might be less able to travel and take part in discussions, study sessions, or training courses. That is why the team travelled each year to various European countries and visited member organizations to offer workshops, lead discussions, share the results of the work done at the organizational level, and motivate local partners to take actions to empower and educate young people. The visits also were an occasion to share campaign materials and maintain EYCE's network. The first dissemination visit took place in Belarus and Ukraine and was used as a preparation for the next initiatives organized in these countries. Furthermore, given the political landscape of Ukraine and Belarus, it was vital to visit them in the year devoted to deepening understanding of the political dimension of ecological justice.

In 2012 the team travelled to Portugal, where they both shared the first results and learned from the local context of nature conservation, biodiversity preservation, and energy saving. The visit also encouraged the local member organization to join the EYCE's Earth Day initiatives in 2012.

The third occasion to call on the member organizations was a project entitled "Sustainability Journey: 4 cities in 4 days!" in which the team travelled to

Latvia, Lithuania, Estonia, and Finland. Members of the team stopped in Riga, Vilnius, Tartu, and Helsinki to meet local youth, raise awareness on environmental issues, and spread the message of the EYCE's Campaign to Promote Ecological Justice. The third visit was the most intense, given the fact that there were plenty of experiences to share. The time was divided between, on the one hand, the more practical and engaging part, such as upcycling workshops, raw-food cooking classes, or production of eco-friendly detergents, and, on the other hand, into more profound discussions concerning the presence of eco-justice in education, international environmental politics, and the role of ecumenical youth in state-based decision-making processes.

Apart from the visits, the dissemination efforts were made through publication of the campaign's magazine *Leaf.* Altogether, six online issues were produced, providing the network of member organizations and external partners with a wide overview of the work being done in the framework of the campaign. The magazine's thematic focus followed the aforementioned central points of interests for each year of the campaign.[7] The member organizations evaluated *Leaf* as one of the most important tools of the campaign through which individuals were able to learn about the activities of the campaign, share experiences from their local contexts, and engage in the eco-justice debate. What is more, *Leaf* offered a place for the guest speakers who participated in the conferences to share their input with a larger audience.

Challenges

Ecumenical youth work today faces numerous problems ranging from financial obstacles in funding the activities to lack of participation. Unfortunately, the Campaign to Promote Ecological Justice is also not immune to these issues. For example, one of the training courses had to be cancelled due to the rejection of a funding application. In addition, in the places chosen for trainings, provision of environmentally friendly venues or vegetarian/vegan catering is difficult to ensure. Furthermore, several online activities did not attract as many participants as was initially expected. Important as well are the changes in the composition of the campaign team. These fluctuations affect the work positively by bringing new energy and ideas on board.

Conclusion

The EYCE's Campaign to Promote Ecological Justice has been a three-year-long process of constant emphasis on the importance of the eco-justice narrative in the ecumenical youth work in Europe and beyond. By targeting both the behaviours and larger sociopolitical landscape, EYCE has managed to widen the theme of care for creation. The overwhelmingly positive response from member organizations was convincing evidence of the need to discuss ecological justice and environmental responsiveness in the Christian context. It could be easily observed that the level of discourse ranged from very down-to-earth perspectives on recycling and water-saving practices to more fundamental reflections about responsibilities as well as theological fundamentals of work related to care for creation.

The work presented in this chapter did not finish once the campaign was concluded. In 2014, EYCE started its next initiative, "Break the Chains! A Campaign to Overcome Poverty." The lessons learned were passed on to the next campaign coordination team, and several thematic focuses identified in the last year of the Campaign to Promote Ecological Justice, such as energy poverty and links between ecological justice and access to resources, could be followed up on during 2014–2016. The first to be implemented was an event entitled "Food, Land and Water: A training course to promote fair distribution of natural resources in the framework of ecological justice."

Finally, the campaign became a venue for personal development not only for those who were participating in the events but also for those who were engaged in designing the whole programme. On one side the campaign helped in approaching the people who otherwise might not be able to familiarize themselves with the ecological-justice framework. But even more importantly, they were enabled to grow as partners in the discussions in their local context of the matters related to creation and well-being of the earth. By adding the justice layer to the programme, we were aware of the fact that we were engaging with an important but also very complex concept. The optimism and courage of the youth yield results larger than one might have expected and, hopefully, they will be a lasting contribution of the youth to the ecological-justice narrative.

CHAPTER 5

Environmental Challenges in Finland and in the Lutheran Church

Ilkka Sipiläinen

Finland is a Nordic country of a little more than five million people and a member of the European Union and the monetary union inside of it. Almost 80 percent of the people belong to the Evangelical Lutheran Church of Finland. It used to be a state church, but is now an independent one. Finland being a wealthy country and the Evangelical Lutheran Church a majority church bring about a big responsibility in environmental as well as justice matters locally and globally. Industrial production in my country has been quite energy intensive, the distances are long, and the climate is cold.

This chapter will describe some of the activities of the Evangelical Lutheran Church of Finland in environmental and climate questions. Those include the the environmental system tailored for the local congregations called the Church Environmental Diploma and the Carbon Calculator, the Climate Programme, the Eco-Fast campaign, as well as advocacy work.

The Church Environmental Diploma and the Carbon Calculator

The "Environmental Diploma" is the cornerstone of the environmental responsibility and advocacy work of the church. The church has had the Environmental Diploma programme since 2001. A local congregation that meets the requirements stated in a 150-page handbook can obtain the diploma. Today more than 100 of the total of 400 congregations have the diploma. All of the biggest congregations like Helsinki, Turku, Tampere, and Oulu have obtained it, which gives a good foundation for the environmental work the church is doing.

It would be hypocritical to speak only about environmental problems and lobby for biodiversity and mitigating climate change if your own house wasn't in order. The project has lasted now more than ten years. The building of the environmental system has been training us all, in the working groups as well as in the local congregations. It has helped networking nationally and locally and has built credibility and interest in the church's environmental work. One example of this is that the church has been granted membership to the National Committee of Sustainable Development chaired by the prime minister or the minister for finance.

The diploma is an official environmental system of the church. It was established by the church council chaired by the archbishop, and every diploma is granted by the very same organ. This is important because we want to integrate it in the very real life of local congregations as a part of social and environmental responsibility as well as good governance. To obtain a diploma the official governing bodies of the congregations have to be committed. The diploma follows other environmental systems in most cases. It has chapters on offices, forests, traffic, targeting carbon neutrality, and so forth, as well as more specifically on Christian environmental education and global responsibility. There is a chapter about graveyards that is very relevant, as the church manages practically all graveyards in the country.

Carbon neutrality has been set as a target for local congregations. In order to help, the church has developed a tailored carbon calculator. One of the requirements to obtain the diploma is that a congregation calculates its CO_2 emissions and sets a target to make them smaller. We have developed this calculator in cooperation with a leading environmental consulting firm and the governmental Finnish Environment Institute.

The Climate Programme

The name of the climate programme of The Evangelical Lutheran Church of Finland is not "Fear, Despair, Stagnation." We wanted to say something that gives added value to the discussion about climate change, something that only a Christian church could and should say. Thus the name of the programme became "Gratitude, Respect, Moderation,"[1] reflecting the theology on which the programme is based. Fear brings about despair that can lead to stagnation. Gratitude for the gift of creation and the love that gratitude generates can make us act for the benefit of nature. The programme has a strong theological and biblical dimension. It gives recommendations both to church bodies as well as church members.

The Origins of the Programme

The idea that the Evangelical Lutheran Church should have a climate programme was born in the general synod in the spring session of 2007. A representative initiative was handled concerning drafting a climate programme for the church. The general synod decided to send the initiative to the church council for preparation and measures.

The church council appointed a working group to prepare a proposal for a climate programme. According to its terms of reference, the programme should (1) give the church as a whole, its congregations, and its church members clear recommendations; (2) arouse discussion on climate issues from the church's point of view; and (3) integrate climate issues into the church's faith, doctrine, and life.

The church council appointed a working group to prepare the programme. In the group there were both theologians and laypeople—the laypeople being, in fact, the proper experts on climate change. Among them were a meteorologist, a member of the European Parliament, a young activist, and environmentalists.

The Theological Basis: Moderation Motivated by Gratitude and Respect

Climate change challenges us to raise a traditional but cast-aside concept of God, according to which God is at the same time present in the creation and beyond it. According to Lutheran theology, it is possible to think that creation is holy in the sense that it is sharing in God's holiness. Creation expresses and mediates God's goodness and love. This way of approaching God's mystery also means that God is nearby, giving us hope that prevents our becoming paralyzed when facing challenges.

God gives all good gifts through creation. Finland is a country full of forests and lakes, and the Finns find nature spiritually important. Many people feel that their relationship with nature is based on gratitude for God's great gifts, a profound sense of unity with all beings, and love for them. This love is a powerful driving force when looking for solutions to climate change and other environmental problems. Experiencing God's presence in nature corresponds to the concepts of Christian theology. Christians gathering together for worship and the holy sacraments as well as those retiring in silence in nature both are part of spiritual life. Nature's healing power is one of God's good gifts. Gratitude gets us to respect nature's own balance in our solutions and to praise our Creator for those gifts.

Human attitudes toward what is holy usually are respect, wonder, and also fear. A certain kind of inviolability is also part of holiness. Understanding the value of creation limits greedy human desire to utilize nature unreasonably. This experience can help us feel deeply the interconnectedness of God's creation. We form a community; we are a part of the web of life.

A moderate lifestyle is an effort to use God's gifts the right way. Today, it means, for example, simplifying life and emphasizing the essential. Human beings give up something good in order to gain something better—more time, space, peace, or possibilities to share one's resources with others. Understood in this way, Christianity is a counterforce to overemphasized consumerism. The life of a Christian is a struggle toward a good and responsible lifestyle, but it is not a joyless life. We can enjoy creation. To do so, we need a moderate lifestyle and ethical choices in consumption. Meditation and prayer give power and direction to ethical action.

Recommendations for the Church Members

It is not very common in a Lutheran church to give any ethical recommendations to the members. Lutheran social ethics is based on the so-called Golden Rule that is written in the heart of every human being in creation.

The climate programme, however, does give recommendations to church members. Mitigating climate change is such an important issue that the church wants to present its members concrete ways to act, "although the church does not have other information about the practical actions than that obtained from the climate change experts."[2] This issue is like that in the age of Enlightenment when the pastors' sermons included instruction on how to grow potatoes. It is suggested that the whole family can gather together to go through the recommendations and to agree on the changes in lifestyle that the family makes. The same can be done by those living alone or in various communities.

It is recommended that the church members assess and change their own lifestyles to be more moderate and study their possibilities in influencing climate-change mitigation. A short list to help in the assessing is given. The programme recommends the periods of fasting preceding Christmas and Easter as a good opportunity to try a simpler and more moderate lifestyle. Also, keeping the sabbath saves the environment by reducing needless consumption. And I already have mentioned the ethical power of meditation and prayer.

In many European countries, like Finland and Germany, there has been pressure to commercialize Sundays. The retail shops have traditionally been closed, but especially the bigger ones have campaigned to change the regulations. That has been called "liberalization"; I could add that we are "liberated" from Sundays. A more proper way would be to call it commercializing them. The churches have been campaigning against the commercialization of Sundays, but in Finland the battle is almost lost and the people are "liberated" from a quiet and different day when society is not running 24/7. It is quite surprising that people even in the environmental movement have understood the Sunday opening as a project of positive liberalization and not a way to put more stress on both nature and workers.

Household emissions are significant sources of greenhouse gases. That is why our own choices matter. People who have made their lifestyles more moderate due to the climate and the environment are encouraging examples of what we can do. The programme states that when your own house is in order, you can demand more of your congregation, municipality, church, country, and the European Union. In that way private changes in one's lifestyle become political acts. They put pressure on political decision-makers as well as legitimate the often-difficult decisions that have to be made for the sake of the well-being of the creation.

There are recommendations on transportation, energy, water conservation, and recycling. An interesting part of the programme contains recommendations on diet. It is well known that production of meat, especially beef, creates a lot of emissions. A related problem is animal rights. Thus it is recommended: "Prefer vegetables, organic food and food produced locally in your diet. If you are an omnivore you can prefer the traditional Mediterranean diet in which meat products are more in a role of seasoning than main course. Reducing the consumption of meat has a great impact on preventing climate change. Even a moderate vegetarianism is a sustainable diet in world scale."[3] The response to the programme has been very positive and there have been already three printings of the Finnish version. The programme is also translated in the other national language, Swedish, as well as in English.[4]

Eco-Fast Campaign

Three years ago the Finnish Environment Institute contacted me and suggested cooperation between our institutions. We went through our various projects and activities and found a shared idea: "eco-fast." Christian churches have been eco-friendly for centuries at fasting times and have great traditions. We started to plan a campaign during the traditional seven-week Lenten fasting time. We wanted to make this campaign ecumenical, and invited the Ecumenical Council to join the campaign.

The target of this campaign is to encourage people to fast before Easter Sunday. The campaign is totally Internet-based and has web pages created by an advertising agency. The outlook is similar to what is found in lifestyle pages, but the content is made jointly by the specialists and researchers of the Finnish Environment Institute as well as the churches.[5] The Institute provides knowledge on how to make your carbon footprint smaller in everyday life. The churches tell about their long fasting traditions and their connection to spiritual life. This kind of cooperation between a governmental research institute and Christian churches is not common in a modern society where many people want to make religion invisible. But it has been working very well.

Advocacy Work and the Theological Message of the Archbishop

All the above-mentioned work is, naturally, building a foundation for the advocacy work the church is doing. I'm not writing about the international and ecumenical part of this work, but will concentrate instead on the national work being done.

I was nominated some years ago as a representative of the Evangelical Lutheran Church to the Finnish National Committee on Sustainable Development (CSD). It is a committee that is nominated by the government of Finland and chaired either by the prime minister or, more recently, by the minister for finance. The CSD helps the government to formulate its policy and is a forum where ministers, leading politicians, civil servants, as well as the representatives of major environmental NGOs, trade unions, and the church meet each other.

Environmental work, climate issues, and global justice have now become a mainline work of the Evangelical Lutheran Church of Finland. The church leaders are generally committed and the local congregations have also committed themselves in the environmental system developed by the church. Archbishop Kari Mäkinen and his predecessor, Jukka Paarma, have been especially active.

Due to secularization and the increasing complexity of society, the church has been losing some of its influence in the Finnish society. It is also increasingly difficult to ring the bells and warn about accelerating climate change. Archbishop Mäkinen, however, managed to do that in his opening speech of the synod in November 2013[6] and got wide media coverage, including a lot of interest in social media.

The archbishop called the researchers of the Intergovernmental Panel on Climate Change (IPCC) a modern version of the prophets of the Old Testament:

> When I read the coolly analytical words of the IPCC scientists, its rumblings seem to urge something deeper than the necessity to urge the development of an action plan, and those rumblings apply no less to the church. I recognize in the report the boldness characteristic of the prophets of the Old Testament. The prophets set the will of God against the prevailing reality and revealed those things that did not conform to God's intention for life. That is what the climate scientists are doing now. Theirs is a prophetic message for our time.[7]

Archbishop Mäkinen was looking for the signs of hope, and reading the climate report he was considering hope from three perspectives. First, hope springs from everyday reality: from the daily round of work and rest, from building for the future, and putting clothes on our children. If these things are lacking there can be no hope or faith in life. We must therefore cling to the continuation of daily life; and all those little things, like the "eco-fast" campaign and the parishes' environmental programmes, strengthen hope. They are on a human scale and create trust in life and for the future. In the same way, hope is maintained by researchers in their work for the development of economically viable, carbon-neutral energy, by companies developing environmental programmes, by decision-makers enacting climate law and formulating strategies, and by the negotiations of national representatives in preparation for the 2015 United Nations climate-change conference. And no less, he added, by those who call relentlessly for justice and the curbing of insatiable economic power. We must give all these efforts determined and unstinting support. Rev. Dr Mäkinen added that keeping such hope is not for Christians alone.

The second source of hope is, according to Archbishop Mäkinen, the hope that is as far removed as possible from optimism and blind faith in progress. "I think of Christ, who is there when all hope is gone and when good works can achieve nothing more. He is there with the victims of climate change. When the challenges faced by the world seem too great, I go in my mind's eye to the crucifix that is closest to my heart."[8]

The archbishop shared this mental picture: "It is a medieval sculpture, from which Christ looks slightly downwards. In his face is an indescribable melancholy and powerlessness—no gentleness, no courage, no great pain, just melancholy. He is the Christ of those who bear the sorrows of the world but who cannot endure them."[9]

Third, there is the hope that this cannot be all that there is. This is the resurrection hope. The world is not a closed space dominated by cold laws of nature and rationally adjustable scientific and economic norms and policies. The imagination—in art and in play—perforates the bounds of a seemingly closed world all the time. And, above all, it is prayer that does this. Prayer is an opening into another world, to whose voice it is attentive. Without prayer, the world would be closed and cold. It is a praying church that keeps hopes alive.[10]

CHAPTER 6

Advocacy for Eco-Justice in Palestine

Simon I. Awad

Palestine is the home to one of the richest biodiversities in the world. Lying at the intersection of three continents, the land has become a crossroads of plants, animals, birds, and people.

At the Environmental Education Center (EEC) of the Evangelical Lutheran Church of Jordan and the Holy Land (ELCJHL), on a pristine hilltop location in Beit Jala's rolling green countryside, it is easy to assume that one is in a country free from environmental problems. This statement is far from the truth, however; a widespread lack of environmental awareness among the Palestinian population has led to environmental degradation, including pollution, loss of biodiversity, and loss of habitat for native animals, which has created conditions that seriously jeopardize the future of Palestine's natural environment.

Continued Israeli military occupation of Palestinian lands has further exacerbated certain environmental problems. Israeli control over Palestinian groundwater resources, severe restriction of Palestinian freedom of movement, the continued confiscation of Palestinian land, and the destruction of countless groves of centuries-old olive trees pose additional grave threats to the region's ecology.

The growing threats related to global climate change, pollution, desertification, and water shortage call for immediate attention and action. Safeguarding the environment and the future of our communities must start now, with concerted efforts to bring about equal access and just resource management. At the EEC, we are committed to the belief that all groups of Palestinian people have the right to a healthy environment, safe drinking water, sanitation, and the right to express their culture and appreciate their environment.

The EEC is working to address these issues of environmental justice through programme offerings for the center's many participants, including students, teachers, women, local graduate students, and decision-makers. Given these programmes, we hope to reinvigorate public attitudes toward the environment, and provide capacity building and awareness training that will empower community members to work proactively toward eco-friendly lifestyle solutions.

Climate and Environmental Challenges

The Palestinian National Environment Strategy has concluded that Palestine feels the impact of climate change in the following ways:

- Rainfall decline and temperature increase are expected, which would aggravate the problem of drought and water scarcity in Palestine. The temperature increase is estimated to range between 2.2–5.1 degrees C and the annual rainfall decline is estimated to be at 10 percent by 2020 and at 20 percent by 2050.[1]
- The climate change is expected to aggravate the problem of land deterioration and desertification, which will compromise the agricultural production and endanger food security.[2] This may have socioeconomic implications in terms of increased poverty and social instability. Internal dislocation and migration to the urban areas and internal disputes on the use of the very limited water resources are good examples of the complications.
- Other possibilities include an increased frequency of natural disasters resulting from drought or extreme climatic events, such as storms, floods, sea-level rise, and heat waves.

The most acute challenge facing the environment in Palestine is lack of water. According to the Palestinian Climate Change Report,[3] the most significant environmental effects of climate change in Palestine in the next century are projected to be a decrease in precipitation and significant warming, thus exacerbating the current lack of available water. The continued lack of water as well as increasing temperatures will cause an increased loss of biodiversity of plants and animals in Palestine. Moreover, food insecurity would increase throughout the country, thus affecting a larger number of people.

Water in Palestine has never been widely plentiful. Of the four climate zones found within the boundaries of historical Palestine, a number of them require only low levels of precipitation. Overall, the country experiences an annual rainy season that the people and the environment rely on for the year's supply of water. Traditionally, Palestinian society has relied on underground aquifers, recharged during winter rains, and above-ground rain collection systems for water. The waters of the Jordan River were also traditionally used for agricultural purposes in the Jordan Valley. Several factors during the past 60 years, however, have depleted many of the natural water sources and contributed to the lack of water security in the West Bank and Gaza.

Most notably, the Israeli occupation has created asymmetrical water distribution and allowed for the pirating of water legally belonging to Palestine. The average daily water use by an illegal Israeli settler residing within the West Bank is more than four times that of a Palestinian. This is in part due to the fact that many settlers come from water-rich countries and continue to act as if there are plentiful water resources. Within the West Bank, Israeli settlements are always guaranteed a continuous supply of water.

Martin Schulz, president of the European Parliament, addressed Israel's Knesset by saying (referring to potable water), "How can it be that Israelis are allowed to use 70 liters of water per day, but Palestinians only 17?"[4] Additionally, much of the water consumed in the Jordan Valley is used for Israeli agriculture. Palestinians are denied access to their water resources from the Jordan Valley simply because of the reduction of flow and contamination of water.

Wells used by Palestinian communities are systematically destroyed throughout the West Bank by the Israeli occupation forces and settlers in an attempt to force Palestinian communities to leave. These actions contribute to widespread water insecurity and also cause water contamination while decreasing the amount of potable water available for use. Intentional improper disposal of wastewater by Israeli settlements has also decreased the sources of available potable water. The problem is aggravated on the part of Palestinian communities since they lack the proper infrastructure (which is denied by the Israelis) for wastewater disposal. Runoff and direct dumping of wastewater has allowed untreated sewage to enter both above-ground and underground water systems. These actions not only cause problems for communities, but also negatively affect local environmental systems, making them unsuitable for some native plant species, allowing for the introduction of invasive and non-native species.

In addition to occupation-related problems, changes to weather patterns have also greatly affected water availability. As yearly precipitation decreases, the amount of water collected above ground has consequently decreased. Thus, the aquifers experience a double impact. As collected rain resources have decreased, more water has been drawn from local aquifers. This has increased the rate of depletion from the aquifers, and now the ancient waters are being removed at alarming rates. Additionally, with decreased precipitation, the aquifers are no longer being replenished at historical rates.

This can be seen most clearly in the Gaza aquifer. The main source of water for Gaza, the aquifer has been overused and is quickly being depleted to the possible point of no return. As more water is being drawn from the aquifer than is being replenished through annual precipitation, the water levels are being dramatically decreased. Due to its proximity to the sea, encroaching saltwater is now being introduced to the aquifer. This water not only is quickly making the water not potable, it is also beginning to deteriorate the structure of the aquifer. The accelerated depletion of all aquifers across historical Palestine is a main indicator of the extremely high level of water insecurity found in modern Palestine today.[5]

In addition to previous factors, other issues contribute negatively to climate change in Palestine. For example, the Israeli occupation's policy of collective punishment against the Palestinian people by uprooting hundreds of thousands of Palestinian trees, the effects of illegal Israeli settlements' industrial zones, the Israeli military areas in the West Bank, and, finally, the wars against Gaza in 2008, 2012, and 2014, with consequent impacts on the environment and climate change.[6]

Another major issue facing the Palestinian environment is an accelerated loss of biodiversity. As the natural water supply is being used up and polluted, the environment is changing in a way that it no longer supports the native variety of plants traditionally found in Palestine. Home to hundreds of endemic species, the protection of this unique area is greatly needed to help stabilize worldwide biodiversity.[7]

Additionally, lack of environmental awareness among Palestinian society entails hunting birds, littering, and food waste. Wastewater issues contribute to a decrease in biodiversity, which increases food insecurity, which has also negative economic effects.

Environmental Education and Outreach

The ELCJHL is dedicated to providing quality education for the future generations of Palestine. In addition to four schools and a kindergarten located throughout the West Bank, the ELCJHL has expanded its educational ministry through the Environmental Education Center (EEC), located in Beit Jala.[8]

The EEC began as a joint project between the ELCJHL and Bir Zeit University. The project, known then as Education for Awareness and Involvement (EAI), was started in 1986 to begin the integration of civil responsibility and environmental issues into school curricula. Focusing on both social and political norms of the region, EAI worked to encourage students to become active citizens in order to strengthen and improve their communities.

Those activities helped form the programme, Children for the Protection of Nature in Palestine (CPNP), in 1992. The creation of this organization formalized the ELCJHL's commitment to the conservation of the environment through environmental education. This organization was based in the schools of the ELCJHL, as well as UNRWA (United Nations Relief and Works Agency), private and public schools throughout the West Bank. CPNP aimed at introducing students to both their Palestinian environment and cultural heritage through the lens of environmental awareness. In 1998, CPNP found a permanent office facility within the grounds of Talitha Kumi School in Beit Jala.

Based in a permanent location and established as a formal education ministry of the ELCJHL, the mission of CPNP was extended to include more aspects of creation care, environmental awareness, and teaching future generations to live environmentally conscious lifestyles as stewards of the earth. A botanical garden highlighting biblical and native plants was built, a natural-history museum created, and a bird monitoring and ringing station constructed as the first of its kind in the Arab world, in order to contribute to the conservation of nature and to contribute to international research about bird migration patterns through the region. CPNP also increased the number of people served to include teachers, women, local graduate students, decision-makers, and interfaith groups in addition to expanding the programmes reaching out to students.

As the work and effects of CPNP continued to grow throughout the West Bank, the organization was renamed the Environmental Education Center (EEC) in 2002. The focus on educating the future generations of citizens and leaders remains central to the work of the EEC, although the aims have expanded to address more issues throughout the communities in the West Bank. Through broad-based educational programmes, the EEC aims at:

- Empowering individuals and communities to take action for sustainable change.
- Supporting the community (particularly students, teachers, women, graduate students, and decision-makers) in developing a deeper understanding of the symbiotic relationship between humans and their environment.
- Promoting local and global awareness of the economic, scientific, and cultural value of specific natural resources, ecosystems, and habitats.
- Providing a dynamic meeting place for students, teachers, and the public.

Since 2002, the EEC's work has been widely recognized not only in Palestine but also worldwide. The EEC focuses on forming strong leaders and programmes within Palestine in order to create a sustainable future. Entering classrooms, providing environmental activities for students and the general public, launching awareness campaigns, hosting festivals that celebrate Palestinian environment and heritage, and helping community leaders to establish sustainable programmes within their local communities are all efforts to create the needed environmental changes within Palestine. The EEC aims to build environmental awareness in Palestinian society. How can you be a Palestinian nationalist without caring for the environment? Additionally, the EEC reaches out to the international community, seeking their support to exert more pressure on Israel to end their occupation of the Palestinian territories. In part, this is accomplished through tours and providing educational materials for international visitors, often in Palestine with church-affiliated groups. The EEC has also hosted five annual Palestinian Environmental Awareness and Education conferences that have been held in Bethlehem, during which local leaders and scientists as well as international guests have both attended and presented papers. Through those conferences and other international conferences attended by members of the EEC staff, further connections are made to strengthen the networking and awareness campaigns established internationally.

The initial work of the EEC focused on reaching out to students specifically. Although the work of the EEC has expanded exponentially, it is still very active within classrooms, student-centered activities, and student environmental clubs. Each semester, educators from the EEC staff visit each of the classrooms of all four ELCJHL schools in the West Bank to give free environmental lessons. Topics include health and wellness, food choices, environmentally friendly actions, resourceful water usage, and waste reduction. During these lectures, the students are encouraged to engage actively in their learning by participating in several activities and are given before-and-after questionnaires to determine how their views about the environment are changing. The EEC also supports over 26 student environmental clubs in governmental and non-governmental schools throughout the West Bank. Students who are participating in the clubs design the year's activity programme with guidance from their supervising teachers and EEC educators. By allowing students to determine the club's activities, students can learn more about what interests them, which allows for deeper and longer-lasting effects on the students, something that can then lead to changes within their communities.

In addition to classroom visits and environmental club activities, the EEC also strives to provide other hands-on events for a broad range of students. Teachers and their students are always welcome to visit the EEC campus and tour the Botanical Garden, Talitha Kumi Bird Monitoring and Ringing Station, Natural History Museum, and Environmental Exhibition. During their visits to the campus, students are exposed to many native plants and a garden set up to accommodate the natural resources present in the environment. Students can see what a sustainable watering system utilizing grey water can look like and how native plants can actually benefit the current environmental crisis, as opposed to the damage caused by introducing new, non-native species. While observing the bird-ringing process, students can watch as many resident and migratory species of birds as possible. This time is also used to discourage students from hunting birds, which is a common practice in Palestine. By showing them the birds up close in the hand, students can learn how important they are for the environment and how to protect them. Students are also introduced to many of the native species found in Palestine in the Natural History Museum. Some of the species are now either endangered or extinct in the region, and students are directly able to observe the devastating effects their actions can have on their environment. Additionally, the Environmental Exhibition informs students about the steps they can take at home and in their communities to work toward a more sustainable future. Projects completed by students from previous summer camps and activities are displayed, as well as educational posters about the Palestinian environment. By continuing to educate students about the importance of the environment and their role within it, the EEC is building strong values in the leaders of the future, helping them to consider the environment within their lives.

Students also participate in many field trips and volunteer activities organized by the EEC. Hikes through several areas are held in order to expose students to outdoor activities and give them a chance to observe scenery of natural environments throughout the West Bank. The most popular hike is through the Wadi Qilt, which runs from Jerusalem to Jericho in its entirety. This hike, sponsored by the EEC, starts in the Jerusalem Desert and ends with a tour of Saint George Monastery outside of Jericho. Students are able to observe a wadi, a valley environment with water, as well as a desert environment surrounding the wadi. For many students, this activity is the first outdoor recreational experience they have participated in, and they enjoy not only the hike, but the bird watching as well.

In order to continue environmental education in the classroom and through environmental clubs, the EEC holds an annual teachers' training. This event is an opportunity to gather teachers concerned with including environmental education in their classroom curricula and provides them with the opportunity to exchange ideas. The EEC provides an expert during each training to help guide the discussions that arise during the conference, but also allows teachers to brainstorm their own ideas to be implemented in their classrooms and communities. The conferences have been well attended for the past five years, and teachers from governmental, UNRWA, and private schools were invited. In a recent training, held in November of 2014, teachers created plans for environmental initiatives within their own communities with the aim of advocating for changes in behaviours and actions in order to work toward sustainability.

In similar work, the EEC has helped establish several women's groups, which focus on environmental actions that women can implement in their homes. The EEC understands that the biggest change at home comes from mothers and wives. By encouraging them to reduce food and other waste, and to reduce water use or encouraging water reuse in safe ways, women can have huge impacts not only in their homes, but also within their communities. The EEC also focuses on encouraging the use of natural or less harmful cleaners, rather than toxic chemical cleaners, and reinforces the environmental lessons learned by students within the schools.

The EEC consistently reaches out to Palestinian society through a variety of events and information campaigns. For the past 14 years, an annual Olive Harvest festival has been held in Bethlehem, and another festival was started four years ago in Ramallah. Additionally, the EEC hosts a spring festival in collaboration with local councils. During these festivals, the Palestinian environment is highlighted and celebrated as a gift to the people of Palestine. Artistic performances and pieces are showcased, often by students from schools in the communities. Local farmers and fair-trade owners are given the opportunity to display their products, which include everything from homemade olives, fresh olive oil, and traditional foods to small handicrafts, new jewelry designs, and other traditional embroidery. The spring festival highlights the creativity of students throughout the West Bank after they have completed a camp held at the EEC. Essays, photos, and paintings created by the students focusing on their understanding of the Palestinian environment are put on display for the public to view. During each festival, winners of each competition are announced and given materials to help them continue pursuing their environmental creative endeavors.

These festivals also open up spaces for all members of Palestinian society to discuss and celebrate their environment. By creating a positive event regarding creation care, the EEC can begin to encourage all members of society to think about the benefits of a clean and healthy environment and opens the opportunity to discuss the impacts their actions have on the world around them. Such discussions are furthered through awareness campaigns that disseminate information through the use of pamphlets, fliers, books, and TV spots.

The EEC has translated the Uppsala Interfaith Climate Manifesto 2008, from the Church of Sweden, into Arabic and distributed thousands of copies to people throughout the West Bank. Representatives of civil society and governmental organizations, political activists, and social workers, as well as Christian and Muslim clergypeople were invited to a workshop where they expressed support to the document and have adopted it. The Uppsala Interfaith Climate Manifesto states that for a long time our planet has been exposed to dangerous changes in the environment, which threaten the future of the earth and human life.[9]

This was brought about during a workshop on climate change, organized by the EEC to discuss the role of religious, political, and scientific institutions as well as the media in Palestine in raising the level of community awareness about this phenomenon. The workshop, under the patronage of the Ret. Rev. Bishop Dr Munib Younan, bishop of the Evangelical Lutheran Church in Jordan and the Holy Land, included speeches by His Grace Bishop Younan, Dr Charlie Haddad, director of education at the Evangelical Lutheran Church; Simon Awad, the executive director of the EEC; Fayez Al Sakka, member of the Palestinian Legislative Council; Rev. Kjell Jonasson, the representative of the Church of Sweden; and Sheikh Barakat, director of the Jerusalem Center for Studies and Islamic Media. Participants included representatives of educational, environmental, and development institutions, academics, political activists, and Christian and Muslim clergypeople.

Throughout the years, the EEC has been able to launch tens of planting campaigns, and we were able to plant thousands of new trees in Palestine, especially in the Bethlehem, Ramallah, and Jerusalem areas. Those trees are a symbol of hope and resilience, creating green space that the public can enjoy and benefit from. Those planting campaigns were particularly important in areas near illegal Israeli settlements and the Apartheid Wall, where Palestinian land is in constant peril of confiscation and infringement. Also, the trees contribute to combating desertification and climate change.

Through these various campaigns, the EEC has continued educating the public about the environment as well as how to advocate for their rights. One step that has been taken is increasing pressure on elected leaders and other decision-makers to place environmental problems higher on their political agendas. The EEC has witnessed some great success with these efforts, which have included visits to the EEC campus by the Minister of Environmental Affairs (now the Palestinian Environmental Quality Authority—PEQA), Dr Yousef Abu Safeyeh. The minister has also signed several memorandums of understanding with the EEC, represented by Bishop Younan of the ELCJHL, to continue collaborating on many environmental issues. In March of 2014, another memorandum was signed for the observance of a Palestinian Environment Day, which would include many events arranged by both the EEC and the PEQA.

Additionally, during the EEC's 2013 Olive Harvest Festival held in Bethlehem, the Palestinian prime minister, Dr Rami Hamdallah, made a commitment to form a committee to work for compensating farmers affected by seizure of their land under the Israeli occupation and to plant 750,000 trees throughout the West Bank.

The EEC also seeks the support of the international community for the continuation of environmental education and for advocacy for eco-justice. Initially, many of the people who were reached had visited the EEC campus for a tour. During their tour, visitors were shown clear examples of violations on the Palestinian environment on the part of the Israelis and were also shown some of the work the EEC is doing to secure a sustainable future for Palestine despite the enormous challenges. Visitors are often prompted to return home and tell others about the problems facing the Palestinian environment and begin advocating for the realization of an eco-just solution in Palestine. By providing visitors with many resources, including some English versions of the pamphlets and books from the EEC annual conferences, the EEC helps equip them for further advocacy.

Many internationals have chosen to further their contribution to the advocacy and education efforts about the Palestinian environment and have participated in the EEC annual conferences. In 2012, the conference was entitled "Eco-Justice of Palestine," during which 16 papers were presented in Arabic, and four were presented in English by international guests.[10] This conference had the highest turn-out from international friends, indicating the growth of international recognition of not only the work of the EEC but, more importantly, the need for sustainable and achievable changes to the Palestinian environment not accomplishable without policy changes both in the policies of Palestine and in ending the Israeli occupation. As more internationals visit the EEC, advocacy campaigns reach out to more people and the message of a need for environmental sustainability and eco-justice in Palestine becomes stronger.

In brief, it became clear that environmental sustainability must be a collaborative commitment from the Palestinian society via several initiatives. Those initiatives have helped educate a wide range of participants about the environmental situation and to equip them with the skills and tools for making a difference. Young people, women, leaders, teachers, and all individuals play a role in preserving and protecting the natural resources of this land. Through our ongoing projects, the EEC hopes to make a lasting cultural impact that will empower Palestinians and promote collective advocacy campaigns for a sustainable change in Palestine.

International Engagement and New Initiatives

In May 2013, in my role as EEC executive director, I attended the World Council of Churches (WCC) Working Group for Climate Change (WGCC) entitled "Caring for Creation Today and Building a Sustainable World for Tomorrow," held in Bossey, Switzerland. During a plenary session entitled "Nature, Religion, and Eco-Justice," I presented some information about the environmental situation in Palestine and the effects of the Israeli occupation. I then described the work the EEC was doing to combat those effects, focusing on education, youth activities, and conferences. Two years ago, the EEC became a member of the WCC-WGCC.

In November 2013, EEC held a *madang* workshop entitled "Eco-Justice For Palestine" at the tenth assembly of the WCC in Busan, South Korea, with two EAPPI (Ecumenical Accompaniment Programme in Palestine and Israel) representatives. Ms. Margit Onstein of Norway and Mr Theo Buss of Switzerland presented their experience about Israeli violation toward nature and humanity in Palestine. In addition, Dr Guillermo Kerber[11] from the WCC gave a presentation entitled "Eco-justice in Palestine: Theological and Ethical Insights," and Mr Nidal Abu Zuloof spoke about Israeli violations against olive trees in Palestine. The workshop addressed the daily environmental injustice faced by the Palestinian people due to the Israeli occupation and called for international solidarity. In addition, I presented a paper on the effects of Israeli occupation against the Palestinian environment. All five presentations increased international awareness and helped form many new connections for possible cooperation in the future.

Representatives of the EEC have also attended the following workshops and conferences:

2014—Germany	The World Council of Churches (WCC) Working Group for Climate Change.
2013—Croatia	Sustainable Education, by Mediterranean coalition.
2012—Netherlands	Eco-justice, Growth and Hope, by European Christian Environmental Network (ECEN).
2011—South Africa	"Youth for Eco-Justice" training, jointly organized by the World Council of Churches (WCC) and the Lutheran World Federation (LWF).

2008—Sweden	Uppsala Interfaith Climate Summit, by Church of Sweden.
2008—Jordan	Climate Change Policies Conference, by United Nation University.
2007—Egypt	New learning for sustainability in the Arab region.
2007—Jordan	Climate Change and Conservation of Marine Environment, a conference organized by the League of Arab States where the EEC was represented by a paper.

Table 6.1 EEC Conference Attendance

The ELCJHL established the EEC in 1986 out of their religious conviction in the need to care for God's creation. They strongly believe in their commitment to environment equity that will not be achieved without real political and economical justice. The church sustains what is stated in Genesis 1:31, "God saw everything that God had made, and it was very good."

In his speech at the third Palestinian Environmental Awareness and Education Conference, entitled "Eco-Justice of Palestine," in 2013, Bishop Younan quoted Psalm 24:1-2, which states: "The earth is the LORD's and all that fills it, the world and all who dwell therein; He founded it upon the seas, and set it firm upon the rivers of the deep."[12]

Also in his speech at the meeting of the Lutheran World Federation (LWF) Council in Indonesia in June 2014, Bishop Younan stated:

> Care for creation is a fundamental commitment growing out of the first article of the Nicene and Apostolic creeds. This is God's creation and human beings are to care for it; we are to tend to God's beautiful garden. There are times, however, when we can distract ourselves with theological language; it is time that the church speaks plainly about the looming crisis of climate change. Since last November, I have become even more convinced that the crisis of climate change is an area beckoning for church involvement in order to provoke societal response.

He continued,

> Trees situated by steadily available water, by contrast, stand up straight, lush and green, a refreshing sign of stability, nourishment and strength for all who encounter them. These are the trees of the streams, the springs, the oases.
>
> The psalmist identifies God's law as the living water that brings such abundant life. When we are guided by the word and will of God, which we know to be most fully revealed in the person of Jesus Christ, we too have this abundant life. This image is found again in Ezekiel 47 and in Revelation 22: "Then the angel showed me the river of the water of life, bright as crystal, flowing from the throne of God and of the Lamb through the middle of the street of the city. On either side of the river is the tree of life with its twelve kinds of fruit, producing its fruit each month; and the leaves of the tree are for the healing of the nations" (22.1–2). Seen in light of this ultimate hope, our theme provides us with an important opportunity to reflect both on the meaning of Baptism and the call to justice.[13]

In light of this, His Grace Bishop Younan has launched many initiatives to care for creation as a church responsibility toward climate change, protection of nature, and conservation of biodiversity through advocacy and actions. These initiatives demonstrate the commitment of Bishop Younan and the ELCJHL to environmental justice in Palestine.

The Creation of Palestinian Environment Day

Resulting from an initiative launched by Bishop Younan, the EEC, in collaboration with the Palestinian Quality Authority, have announced that the fifth of March will be dedicated as Palestinian Environment Day. Palestinian Environment Day will be celebrated annually with a focus on raising awareness about the environmental situation in Palestine. This day aims at fighting for environmental rights, for sustainability of the environment, and for showing both locals and internationals the beauty of the Palestinian landscape.14

Caring for Birds

Bishop Younan announced during the third Palestinian Environmental Awareness and Education Conference the launch of the first four National Bird-Watching and -Ringing Weeks in Palestine. These weeks were held in collaboration with the Palestinian Environmental Quality Authority in an attempt to draw attention to the treasure of biodiversity in Palestine and for more protection of it. Hundreds of students, journalists, volunteers, and researchers participated in the activities. Those activities were held at the Talitha Kumi Bird Monitoring and Ringing Station, Al-Makhrour, Tulkarem, and the Jerusalem Desert of the Wadi Qelt in Jericho.

The EEC appreciated the role of Bishop Younan, as the head of the ELCJHL, for his support of environmental awareness in Palestine and the importance of teaching children how to care for creation, pointing out that Palestinian society has always cared about nature, calling on the clergy and educators to urge citizens to increase their interest in the environment.

Bishop Younan also launched an initiative at the EEC's conference in 2012 to officially adopt the Palestine Sunbird as the national bird of Palestine, in collaboration with the EEC and Palestinian Environment Quality Authority. After years of hard work, the Palestinian Cabinet of Ministers officially adopted the Palestine Sunbird as the national bird in February 2015.

Bishop Younan Environmental Award

His Grace Bishop Younan has established an annual environmental award in his name, to be presented each year on Palestinian Environment Day. The award celebrates distinguished students, researchers, academics, and community leaders in Palestinian society who are committed to raising environmental awareness and working for the protection of the environment in Palestine.

As already noted, two of these four initiatives have been officially adopted by the Palestinian Authority as a result of collaborative work by the EEC and the Palestinian Environmental Quality Authority. The creation of Palestinian Environment Day as an official national day and the Palestine Sunbird as the national bird in February 2015 are remarkable accomplishments that demonstrate the importance of the environment in Palestinian society and the commitment of the Palestinian Authority to continue to work for environmental conservation. As part of the first Palestinian Environment Day organized by the EEC and the Environmental Quality Authority in 2014, more than 2,000 students and community members participated in activities throughout the month following 5 March.

In the future, the EEC will work on five strategies:

1. To raise environmental awareness and help develop capacities.
2. To help protect natural and cultural heritage and encourage eco-tourism.
3. To help create broad-based grassroots movements to contribute to peace building, reconciliation, and eco-justice.
4. To advocate for the Palestinian people's environmental rights and help them become environmentally responsible.
5. To help create capable and responsible citizens and strengthen Palestinian identity as well as a sense of civic responsibility.

These five strategies will be implemented through:

- Raising awareness regarding important environmental issues, such as climate change, global warming, biodiversity, pollution, natural resources, eco-tourism, and environmental enhancement by making classroom visits and hosting campaigns, workshops, and seminars.
- Improving public awareness about biodiversity and natural resources, and promoting eco-tourism in Palestine through development of the EEC campus.
- Protecting local biodiversity by planting additional native trees and plants and developing the EEC's Botanical Garden into a "Biblical Garden."
- Raising awareness regarding environmental challenges in Palestine and the obstacles and consequences of Israeli military occupation.
- Improving advocacy for sustainable resource management and economic development on peace building and eco-justice.
- Fostering a sense of civic responsibility through hands-on engagement and volunteerism.
- Developing understanding among students and educators about active citizenship and instilling national pride, as well as relinking the Palestinian identity to the land through environmental education, culture, and history.

The EEC is planning to launch advocacy and lobbying campaigns to incite international communities to pressure Israel to end its occupation and respect international conventions relevant to the environment. It will also call for the need to respect laws and international conventions in relation to the issue of global warming and the resulting climate change, seeing it as a duty for everyone to preserve the environment for the rights of future generations. The EEC will call for environmental responsibility of individuals and groups in Palestine and around the world, in order to contribute to reducing the impact of climate change and to protect the planet.

The EEC will call for an alliance with leaders of other faiths, locally, regionally, and globally, especially in the Middle East, to raise awareness about the great danger that threatens humanity and to encourage them to take responsibility together. The EEC notes that shared religious values converge on this humanitarian objective, that is, to preserve the greatness of God's creation, to mobilize citizens' initiative to protect the earth, to replace fear with hope, and begin on the path of sustainable production and consumption. Finally, the EEC will call for working according to the principles of justice, solidarity, and empathy to encourage scientific research and help create the best leaders to protect the earth.

CHAPTER 7

One Earth, One Sea, One Sky

Joy Kennedy

Earth Is Our Home

Earth is the only home we have in the vastness of the cosmos.[1] It is a sacred gift from the Creator. It is time for us to wake up to what is happening in our home.

The scriptures are replete with references to the marvelous intricacies and interconnectedness of all creation, including our human situation within it. The struggle to understand the very meaning of our existence and our relationship to the Creator and the universe is as vital today as it was recorded in biblical times. But perhaps now there is an even more eschatological dimension as we face the impending threat of drastic changes to earth, its land, its ocean, and its atmosphere, for which we humans have had a disproportionate responsibility in recent years. We as a species have demonstrably acted as the agents of disruption and destruction to the only place we call home. Can we learn to care and act to help bring about the transformation envisaged in the foretold "new creation"?

The ancient psalmist was succinct when he exclaimed: "The earth is the LORD's and all that is in it, the world, and those who live in it; for he has founded it on the seas, and established it on the rivers" (Ps. 24:1-2).

God the Creator asks of a struggling Job: "Where were you when I laid the foundation of the earth? Tell me, if you have understanding. . . . Or who shut in the sea with doors when it burst out from the womb? When I made the clouds its garment, and thick darkness its swaddling band, and prescribed bounds for it" (Job 38:2, 5, 9-10).

But humans continue on a business-as-usual course, causing St Paul to admonish: "Besides this, you know what time it is, how it is now the moment for you to wake from sleep. For salvation is nearer to us now than when we became believers" (Rom. 13:11). The night is far spent, and the dawn is at hand. Perhaps it is not too late.

One Earth

We are one earth community upon this tiny blue planet, orbiting round one rather small star, interconnected and interdependent on one earth, one sea, one sky—God's wondrous gifts for life. We dwell together with all living things in this home, sometimes known as Gaia. How then shall we live? What is the reality today? What lies ahead in the future? We face a challenge to reexamine and understand afresh our relationship with creation for our very survival. One land, terra firma, is where we mostly live, surrounded by one global ocean, under one atmosphere. Our Creator God calls us into *covenant with all creation*, into communion and community in the commons of creation. Our part of the *rainbow covenant* is to love this small planet we call home and bring harmony in the earth community.

Reflecting on the image we have of ourselves in this relationship leads us to the need for a broader, more holistic, and more inclusive theology, a theology of whole-earth justice. Jesus of Nazareth declared, "I came that they may have life more abundantly"—not "wealth," or "stuff," or "labour-saving devices"—but LIFE (John 10:10). That includes clean air to breathe, water to drink, enough food, and sustainable community, peace, and harmony. Enough for *all*, to be shared by and for all species and all generations. That is part of the *new covenant* with all creation Jesus wrought through his death and resurrection.

Proclaiming such a theological frame calls for some new or reworked concepts such as a theology of home and a theology of place and of land, and an emphasis again on our relationship to the Creator and the creation. If, as I was taught, I am my hymnody and my psalmody, then I am stirred to hear Isaac Watt's classic hymn paraphrasing Psalm 90:

> O God, our help in ages past, Our hope for years to come,
> Our shelter from the stormy blast, And our Eternal Home . . .
> Before the hills in order stood, Or Earth received her frame,
> From everlasting, Thou art God, Through endless years the same.

But where and what is home? If home is where the heart is, what stories do we tell about it as we try to understand our changing relationship to our home planet? We need to be about seeking a new narrative that can lead us to a new ethic of home and changed institutions, behaviours, and actions that are required in our time.

In his classic work, *A Sand County Almanac*, Aldo Leopold says: "It is inconceivable to me that an ethical relation to land can exist without love, respect, and admiration for land."[2] This could perhaps be expanded to include the whole of creation as humans seek to better express their place in it.

Walter Brueggemann helps us to see that land is never simply physical dirt but is always freighted with social meanings derived from historical experience.[3] The yearning for land is always a serious historical enterprise concerned with historical power and belonging. Our humanness is about historical placement in the earth with excess meanings rooted in and moving beyond literalism.

Land is a central, if not *the central theme* of biblical faith. Biblical faith is a pursuit of historical belonging that includes a sense of destiny derived from such belonging.

How do you think of "home"? What kind of matrix of words and images conjure up your experience and longing for that place you identify as "home"? These will reveal much about how you engage with it in your behaviour and life practices—and similarly thereby the impact of each individual and community upon the earth itself. It matters greatly as it contributes to the formation of our belief system—our worldview—for as we truly believe, so we act.

Our Home the Earth Is in Crisis! The Global Ocean Is in Crisis! The Atmosphere Is in Crisis!

What is going on? How do we get out of the mess we are in? Jesus warned, "You cannot serve both God and wealth" (Luke 16:13). The word *economics*, like "ecology," and "ecumenism," comes from the root Greek word *oikos*, meaning "house/household" and is about global housekeeping. Our planetary living room needs some serious global housecleaning!

The global financial crisis is an opportunity which is in danger of being missed. Originating in the richest parts of the world, it points to the immorality of a system that glorifies money and has a dehumanizing effect by encouraging acquisitive individualism. The resulting greed-based culture impoverishes human life, erodes the moral and ecological fabric of human civilization, and intoxicates our psyche with materialism. It needs cleaning up.

There is growing awareness of our common vulnerability and the limits of our current way of life. Those most affected—who should be listened to as modern-day prophets—are women, who bear a disproportionate share of the burden; young people and children, as their doubts are raised and their sense of security for the future is eroded; and those living in poverty, whose suffering deepens.

The crisis we face is at the same time systemic, moral, and spiritual.

One Sea

The global ocean makes up 99 percent of the living space on the planet—because of its immense depth. It covers 70 percent of earth's surface. All the land surface and narrow atmosphere where life survives equals only 1 percent of the inhabited planet.

In her book *Sea Sick*,[4] Alanna Mitchell, one of the first journalists to do so, documents the growing evidence of the crisis in the global ocean. Why is the ocean so important? She describes how the ocean controls climate and temperature and the carbon and oxygen cycles of the planet, as well as other chemical systems that give all living creatures life—including us.

To understand the important connection between the ocean and climate, we must know a few facts. First, 33 percent of the ancient carbon dioxide (CO_2) in the atmosphere from the burning of fossils has been absorbed into the ocean. Second, 80 percent of the extra heat held against earth from the extra CO_2 has been absorbed by the ocean. It is now estimated that the global ocean temperature is up 0.33 degrees Celsius from preindustrial times, an average across the ocean and much higher in parts of it.

The amount of carbon concentrated in the atmosphere varies over time, but scientists warn that 350 parts per million carbon dioxide molecules or equivalency in the atmosphere (ppm by volume) is the maximum level of CO_2 to avoid accelerated climate change. Today we are at *400 ppm* and rising, whereas 250 years ago it was at 250 ppm. The United Nations negotiation target is for up to 450 ppm, resulting in a two-degree Celsius global average mean temperature increase over preindustrial levels. But the most vulnerable UN member states are arguing for the accepted limit of 350 ppm, as they are already seeing climate impacts from rising seas, intense typhoons/cyclones, heatwaves, floods, and droughts. The negotiations for a fair, ambitious, and binding agreement on climate change have been extremely complex and difficult. But there continues to be hope that leaders will take action for climate justice.

The Union of Concerned Scientists says that the excess carbon that you and I have been dumping into the atmosphere for the last half century—from deforestation and from burning coal, oil, and natural gas—has pushed CO_2 levels more than 25 percent higher than when measurements of air began at Mauna Loa Observatory in Hawaii in 1958, when levels were at 317 ppm, according to Scripps Institute of Oceanography[5]. Each year, the amount of CO_2 in the atmosphere rises during the northern-hemisphere autumn and winter, then decreases through the spring and summer as plant growth removes the CO_2 from the air during the growing season. The annual peak usually occurs in May, but in 2014 we experienced this new milestone of 400 ppm CO_2 for a couple of months before the annual decline occured.

In fact, the level of carbon dioxide in the atmosphere reached 400 ppm for the first known time in human history[6] in May 2013, and again on 20 March 2014, two months earlier. It broke a global benchmark for the whole month in March 2015, and it will stay above 400 ppm for longer. At that rate, it will only be a handful of years until we are living in an atmosphere permanently above 400 ppm, which is a somewhat arbitrary marker, as humans did not exist the last time atmospheric CO_2 was at that level.

Linked to atmospheric carbon concentration is its concentration in the ocean. Some more facts are necessary to understand the connection between atmospheric carbon and oceanic carbon. Carbon absorbed from the atmosphere by the ocean is called "blue carbon." More carbon is absorbed by the ocean than by trees, plants, and soil and it is often seen as the biggest "carbon sink." But how long can this be taken for granted before irreversible changes appear? One of the critical questions now being researched is how carbon saturated the ocean is becoming.

One of the emerging observations is that the chemistry of the global ocean is changing: in fact, ocean acidification is well underway, as carbon dioxide absorbed by water becomes more acidic: CO_2+ H_2O = carbonic acid. The indicator is the ocean's pH balance.[7] The changes are measured in the pH levels on a scale from (acid) 1 to 14 (base). The pH of water is 7, and for human bodies it ranges from 7.35 to 7.45. Any change has severe implications for the survival of species. Five global extinctions in world history were due to changes in the oceans' chemistry.

One of the most devastating impacts being observed today is on corals and, in fact, all marine life which is vulnerable to acidification. Corals are the nursery for most fish. Acid bleaching kills them and fish habitat. The increased carbonic acid has a deleterious effect on the creation of shells and skeletons of marine creatures, which are mostly made of calcium carbonate. Will they survive? Will we survive without them?

Another change that is affecting marine well-being is the impact of rising sea levels resulting from thermal expansion of the water as it warms and from melting of land-based ice and glaciers such as the Greenland Ice Sheet and Antarctic Ice Shelf. Estimates vary, but could be as much as 18 to 59 centimeters (7.1 to 23 inches), according to a 2007 report from the Intergovernmental Panel on Climate Change (IPCC), or higher as more recent studies indicate, such as the 2013 IPPC Assessment Report AR5, which projects sea-level rise of 1 to 3 meters per degree of warming if the warming is sustained. This, of course, creates real concerns for low-lying islands and coastline communities.

Added to that, global warming has accelerated—more in the past 15 years than the prior 15 year—because 90 percent of overall global warming heats the oceans, and the oceans have been warming dramatically. The "missing heat" has been found in the deep oceans. Thirty percent of ocean warming over the past decade occurs in the deeper oceans, at depths below 700 meters, unprecedented over the past 50 years. Some recent studies[8] conclude, based on slowed global surface warming, that the sensitivity of the climate to the increased greenhouse effect is actually lower than the IPCC best estimates. The fundamental flaw is that they do not account for the warming of the deep oceans. Slowed surface air warming has lulled many into a false unwarranted sense of security.

The combined results of all these effects are becoming increasingly visible. Observations point graphically to links between climate and ocean change in the ocean's dead zones, where there is no oxygen and little life. The count so far is more than 407 known dead zones in the global ocean; for instance, one is 20,000 square kilometers in size. Some are due to pollution, such as nitrogen fertilizer run-off, creating algae blooms. At least four are directly due to climate change. Oxygen depletion equals less life.

What happens when your garbage ends up in the ocean? Marine litter is mushrooming as many discarded waste products are toxic and nonbiodegradable. Heavy metals and pharmaceuticals are bad enough, but the most common and insidious form is plastic of all types. But invisible microscopic plastic molecules are the most invidious of all. This adds to the dead zones as well, making it increasingly hazardous for marine life.

The end result of such deteriorating ocean habitat is the dramatic rise in extinctions. It is interrelated to carbon concentration in the atmosphere and ocean acidification and warming. CO_2 levels have been as high as 1900 ppm in earth's past. The "Permian" extinction (252 million years ago) was called the "Great Dying": 95 percent of species became extinct due to a massive eruption of CO_2. At least 20 percent of extinctions are happening today—maybe up to 70 percent by 2065. Scientists name it "planetary destabilization" and we humans are intimately involved in the process.

So how does it all add up and where are things headed? One economic means of calculating the cumulative impact is in the following study:[9]

> Date: Fri, Jun 6, 2014 at 12:56 PM
>
> *Subject: Seas store 500 megatonnes of atmospheric carbon every year*
>
> A study published today has revealed the extent to which life in the high seas is mitigating climate change, taking up a staggering 500 million tonnes of carbon per year by storing one-and-a half billion tonnes of carbon dioxide away from the atmosphere.
>
> Filling one of the gaps in knowledge identified by the last IPCC, it is the first study to assess the ecosystem services of the high seas and place an economic value on them. Commissioned by the Global Ocean Commission, it identifies 15 ecosystem services of direct value to humans ranging from 'provisioning' services such as genetic resources and raw materials, 'regulating'

such as air purification and biological control, to 'habitat' services such as life cycle maintenance and gene pool protection.

The study calculated an economic value for the role of high seas carbon sequestration as between US$74 and US$222 billion annually.

Another way to count the cost is in the loss of fish. Global populations of *all* large predatory fish fell by 90 percent within 50 years after industrialized fishing began. Reasons for collapsing fisheries include: pollution; overfishing; ocean warming; oxygen depletion; ocean acidification; coral bleaching and die-off; discarding caught nonpreferred species. The extinctions are rapidly increasing as these causes are compounded.

One Sky

We all live in the biosphere under the one enveloping protective shield of the atmosphere and the stratosphere. If, indeed, there is a time for every purpose under heaven, then we must also acknowledge that what goes on in the noosphere of human thought can—and is—making a difference. Climate and ocean change is dynamic. This essay attempts to describe a brief overview of many of the interconnections of known factors as they develop into the future—acknowledging there are many unknowns. With so many interlocking factors, much of the interconnected data is yet to be analyzed and explained, but much is becoming clearer as time passes. This complicated picture has many more aspects of course, as all natural systems do.

What is clear, however, is that the timeline from preindustrial to industrial to the present age, and projecting into the future, has shown an accelerating increase of CO_2 concentration in the atmosphere and ocean (averaged across all zones and more highly concentrated in some regions at different times). It has shown change in pH levels, an increase in ocean acidification, and increasing extinctions of marine and land species.

One particularly dangerous effect is that on the health of plankton, responsible for half of the global production of oxygen. Another way to look at it is: every second breath is brought to you by plankton.

Geologists are naming our era "the anthropocene"—because we humans are having such an impact on the life and chemistry of our planet. We do not know how close is the tipping point. We have an ethical and moral obligation to take responsibility and act now for climate justice, because, as my favourite posters say, "Nature Doesn't Compromise" and "There Is No Planet B."

Climate and ocean change is a moral, ethical, and spiritual issue. As such, we can take heart from evidence that the world community has come together to address other moral and spiritual issues with integrity and courage, and prayer and concerted action that has led to positive change—much of it still ongoing. To name a few, movements against slavery, child labour, acid rain, and for civil rights, women's rights, human rights, and ecological rights have all made strong gains and give reason to hope.

Will we do so now in the face of the global climate crisis? How?

Eco-Justice: Visioning a "New Normal"

Then I saw a new heaven and a new earth. . . .
Behold, I make all things new.
—Rev. 21:1, 5

What is "normal"? We live in a socially constructed world that interacts with the natural environment—the creation of a loving God. Human systems and their norms are constructed on underlying belief systems, or ideologies, with values at the core. Beliefs and values change societies—but also change as societies evolve. Systems created by humans can be and constantly are recreated by humans. Think of examples in your own experience.

What is held as the highest value? What do humans value most—and worship: the market and money, or community and relationships—and a sustainable planet? When the dominant economic principle is greed and self-interest, the result is widening disparity and ecological destruction—then we can feel the great cry for transformation and the revelation for which "all creation is groaning" (Rom. 8:19). Instead, a new paradigm and norms are needed based on principles of equity, fairness, mutuality, justice, and sustainability. As a statement from the United Church of Canada declares, "In and with God, We can direct our lives toward right relationship, with each other and with God. We can discover our place as one strand in the web of life."[10]

All life on this small planet is interdependent and interconnected. Air, water, land, and energy sources provide the necessary environment to sustain life. But the structures created by humans for living within the limits of the earth's ecosystems and its carrying capacity are stretched and fragile. A renewed theology of earth as God's sanctuary helps to break the vicious grip of a belief in the prevailing paradigm of growth at all cost and of unfettered economic globalization. We can learn to move into a new way of being earth community in this time and earth-place, living out God's mission, being the change we wish to see.

We know that poverty, exclusion, wealth concentration, financial crises, climate change, pollution, and ecological destruction require moral, ethical, and spiritual solutions. Walter Wink describes the present world as ruled by the "Domination System" for which "Christians are to work to prepare the conditions for God's Domination-free Order (1 John 2:15-17)." Those liberated from the tyranny of the old order receive a new, holy spirit—"not the spirit of the Domination System (*kosmos*), but the Spirit that is from God (1 Cor. 2:12): the Spirit of truth, which the Domination System is not able to receive, because it can neither recognize it nor comprehend it. You know it, because it is already in your midst, and will be inside your very beings (John 14:17). This new Spirit re-socializes believers into the New Reality."[11]

Seeking a New Narrative

We are not alone and can operate with faith in the future and our ability to change our behaviour and social norms. We envision eco-justice as a higher principle and claim an ancient vision of the "new earth." New choices must be made—choices to uphold sustainable communities and an economy glorifying life, to include: relationships, care and compassion, solidarity and love; aesthetics and the ethics of life, participation, and celebration, cultural diversity and community vitality, responsible growth that recognizes human responsibility for creation and for future generations.[12]

Re-visioning and reconstruction requires a reconceptualization of wealth beyond money that supports sustainable communities and acknowledges its ecological debt.[13] For eco-justice to become the new normal, to restore the sacred balance, and repair the breach, we must replace:

- Competitive globalization *with* multilateral cooperation.
- Commodification of the global commons *with* conservation of the sacred gift of the earth.
- Economy based on greed *with* economy of enough.

The gospel is about the coming of the new age, "on earth as in heaven." Resurrection is the hope and the invitation to transform our global society to live in unity, solidarity, and sustainability; with equity, harmony, compassion, and love. God our Creator, Redeemer, and Sustainer is calling us to a new way of being and living as one earth community. Are we listening?

American theologian Barbara Rossing gives this cogent reflection:

> The book of Revelation, the Apocalypse in which John of Patmos is writing to seven churches who have been held captive, their imaginations seduced, is a 22-chapter long journey of helping to pull back the curtain to expose the empire of Rome for what it is, which is an evil beast, but also a very seductive presence that is very tempting and alluring in many ways, that it offers an economy that seems to make promises. But John says they are false promises. This is a book about structural issues, political economy, the two contrasting cities at the end: Babylon and New Jerusalem. This is an incredible journey of liberation with the invitation to come out of empire; it is an exodus journey out of participation in that unjust political, economic system that is the Roman imperial system. He also lists the lamentation of the merchants, and the kings and the sea farers, all of those whom are losing their cargo, and their trading economy, they are all lamenting because the structures are changing.
>
> John has the vision to see that already God's new political economy is dawning in our midst, and the whole book is this amazing journey of hope out of captivity to that system that has hypnotized people. And then we go on this exodus journey into God's alternative vision which is the New Jerusalem vision, where there is a river of the water of life that is given free of charge. You don't have to participate in the money system, or be rich, to be able to drink of God's river of life. The tree of life bears fruit and its leaves are for the healing of the nations. This is a vision that we can still draw on without minimizing how terrible the crisis is. The pathology of empire is so deep and it's in people's heads, it's everywhere; they've been seduced by it. But he is drawing us out to see a different political economy that is not just a vision of heaven after we die, but it is God's alternative political economy that we can live in today, and it is structural.[14]

At times much of this information and analysis can be heavy, overwhelming, and even frightening, and it is important for us to connect to it in a personal and faith perspective. *O God, have mercy on me—the sea is so wide and my boat is so small.*

We Are Not Alone

Christian faith encourages us to act responsibly and together with courage, with the inspiration of the Holy Spirit and in the love of Christ for the healing of the world. We are all in the same boat, our earthship afloat on the sea of grace. The imposing sculpture by Aboriginal Haida artist Bill Reid that dominates the courtyard of the Canadian Embassy in Washington, D.C., is called *The Spirit of Haida Gwai.* It represents all lifeforms, including humans, in a magnificently carved canoe headed somewhere—not knowing where, but with hope and faith.

According to Keith Helmuth of the Religious Society of Friends, "This commonwealth of life, this voyage of the great canoe is a sacred journey. Solidarity is the key. Right relationship is the guide. Economic adaptation that enhances life's commonwealth and advances the sacred journey is our moral assignment and ethical responsibility. It seems reasonable to think this is what the Creator must have in mind."[15] We are a part of the one earth community. As the United Church of Canada has declared, "Divine Creation does not cease until all things have found wholeness. . . . In the meantime, we embrace the present . . . caring for the Earth, Choosing life."[16]

Ross L. Smillie's book, *Practicing Reverence*, which promotes an ethic for sustainable earth communities, is a good place to start for ideas on becoming good earth citizens and, in his words, faithful "stewards of the sacred." As he writes, "The shift from destructive to sustainable patterns of living is the great work of our time. In the ancient words of the Christian tradition, it is *metanoia* (repentance) and *conversio* (conversion), both of which mean literally to turn around and change direction. They are about more than just thinking differently, but living differently, even being different."[17]

Living Hope

How can we live beyond despair and into hope? Here are a few practices to get started:

- Offer a prayer of gratitude and thanksgiving for the precious gift of earth—and your small part of it.
- Breathe deeply and feel your lungs fill up. Exhale a prayer of thanksgiving for the sustaining gift of air.
- Reflect on the gases that make up the atmosphere and how important it is that they stay in balance. Reflect on the ocean's role in that, especially plankton, as you humbly accept the gift of oxygen.
- Practice hope daily. Fast and pray for the climate.
- Recommit to caring about the future and caring for home.

Concluding Thoughts

As we discern what God requires of us today we can look at the evidence all around us, project into the emerging future, reflect back upon wisdom from the past, and decide on a path for living in relationship with this earth community. In deciding for a just and sustainable vision, we must ask what we can do individually and together in our churches and community organizations to foster an ethic and culture of eco-justice where our actions match our beliefs and intentions. We can practice hope with integrity and sing a new song of love and gratitude as we make changes in lifestyle, behaviour, and political and organizational structures. Let us love this planet, pay our ecological debts, and join together in the building of a sustainable, loving, compassionate, resilient earth community, acting and advocating for the changes that will help to preserve and protect this beautiful and fragile earth, our island family home in the universe.

May God grant us the wisdom, grace, and courage to do so.

Part Two

Eco-Theology and Climate Justice

CHAPTER 8

Religious Responses to Climate Change

Turning Faith into Action

Elias Crisostomo Abramides

Blessed are the peacemakers,
for they will be called children of God
—Matt. 5:9

Climate Change and the World Council of Churches (WCC)

In the Nicene-Constantinopolitan Creed we read: "We believe in one God, the Father Almighty, Maker of heaven and earth, and of all things visible and invisible."[1] God is the maker of all life, which is God's free gift of grace. Life in all its forms is a grace, a precious gift given to us by the Creator, and our most sacred custody. Therefore, it deserves our respect, our devotion, our reverence, and our care.

As Christians, but moreover as people of faith, the imperative question we have to formulate for ourselves at this stage has to do with what we have done, what we are doing, and what we will do to protect God's good creation.

Climate change is a threat to God's creation, occasioned by the emission of anthropogenic polluting gases that are introduced into the atmosphere. The pollution of the atmosphere, its consequential warming at a global scale, and, finally, the climate change associated to that, are one of the greatest threats challenging the existence and continuity of life on planet earth, our home. It puts at risk hundreds of millions of people today and billions more in the future. No part of creation will escape the impact of climate change.

It is because of the World Council of Churches' universal scope—or "catholicity"—that the WCC and its member churches are concerned with climate change and involved in its study for common advocacy and actions addressing its origins, impacts, and consequences. This has been the main reason for our presence in different international meetings and in the yearly Sessions of the Conference of the Parties (COP) to the United Nations Framework Convention on Climate Change (UNFCCC). The main interest of the WCC during the international negotiations is to be the voice of the voiceless, including the future generations who will inherit our wise or poor decisions. We need to call the negotiators' attention to issues of justice, ethics, morality, solidarity, and equity. Without these fundamental concepts no durable and sustainable agreements will be achieved. Moreover, the WCC has stated that climate change hits basic human rights, specifically the right to life, the right to water, the right to food, the right to health, and the right to housing.[2]

The WCC has centred its attention on climate change since 1988. We remember the names of the pioneers and followers who struggled to create awareness on this issue among the churches, and among the members of the global civil society, creating fellowship among the great world religions: Lukas Vischer, Wesley Granberg-Michaelson, David G. Hallman, Konrad Raiser, Samuel Kobia, Martin Robra, and presently Olav Fykse Tveit and Guillermo Kerber are special landmarks in this worldwide saga.

The founding of the WCC Working Group on Climate Change, which took place during the United Nations Conference on Environment and Development (UNCED),[3] is a unique group within the WCC, with a distinctive composition. The members meet once a year as a group and participate in the UNFCCC yearly activities around the world where the UN Conferences on Climate Change are convened.

It is important to denounce the aggressions against creation; however, much more is at stake. Respect for creation and human beings as God's children is the indispensable precondition for a meaningful struggle against environmental degradation and climate change.

We need to recognize and accept the intimate ethical and deeply religious implications of climate change. It is a matter of justice, it is a matter of ethics, of solidarity and equity, and it is a matter of love: love for God the Almighty, love for our neighbour, and love for creation.

The prophet Micah asks: "What does the LORD require of you but to do justice, and to love kindness, and walk humbly with your God?" (6:8). The gifts of God's creation must be fairly shared. We are living on a finite world with finite resources and within given planetary boundaries, which obviously indicates that infinite growth in consumption of energy and natural resources is impossible in a finite system.

By ignoring or denying the warnings of the scientific community, world leaders and policy makers consciously decide to accept the destruction that is likely to result from climate change: people dying or being displaced due to storms, hurricanes, and floods, people losing their habitat, people being exposed to new menaces, and losing confidence in the future.

Just a few weeks before the 19th session of the Conference of the Parties to the UNFCCC in Warsaw in 2013, the leading climate scientists of the Intergovernmental Panel on Climate Change (IPCC) for the first time unfolded how much more carbon dioxide humans can introduce into the atmosphere without triggering critical levels of climate change. In order to limit the atmosphere's warming caused by anthropogenic CO_2 emissions to two degrees Celsius above preindustrial levels, with a probability of 66 percent, the cumulative CO_2 emissions (expressed as gigatons of carbon: GtC) from all anthropogenic sources is required to stay between zero and 1,000 GtC. More than half this amount had already been emitted by 2011.[4]

Taking into account the undeniable weight of evidence, it is incomprehensible why there are so many involved in this crisis who are still dubious about achieving the very urgent coordinated action needed to reduce greenhouse-gas emissions. Every year we are a step closer to the tipping point at which scientific studies warn us that climate change will become irreversible.

Climate change challenges the global ecumenical movement to become even more explicit about our own vision of the *oikoumene*, the entire inhabited earth as God's common household. As the WCC clearly emphasizes in its statement to the High Level Ministerial Segment of the 19th COP:

> Life in all its forms must be kept enjoying the perfect natural equilibrium of the natural world, which has its own laws and rhythm. Contemporary life is menaced by factors unknown in past centuries: today we speak of food insecurity, of water insecurity, of energy insecurity. Is this the scenario we are preparing for the generations to come? Is this the world we shall leave to our descendants? Are we so isolated from the present reality, from the menace of peoples whose territory will disappear under the oceans' waters, whose habitat will be not only jeopardised but ultimately destroyed by the advancing desertification? Shall all the remaining species disappear leaving the world we live in without its rich and life-sustaining biodiversity?[5]

The Ecumenical Patriarch's Involvement

The Ecumenical Patriarchate is an Eastern Orthodox Church that has fully participated in the life and activities of the WCC since its inception in 1948 and is one of its founding members. On 2 November 1991, His All Holiness Bartholomew was enthroned as the 270th Archbishop of Constantinople-New Rome, and Ecumenical Patriarch. He had earlier served on the WCC executive and central committees and in the WCC Faith and Order Commission, has worked to advance reconciliation with the Roman Catholic Church, the Anglican Communion, and other Christian confessions through theological dialogue, and joins in dialogue with the monotheistic faiths of Islam and Judaism. Ecumenical Patriarch Bartholomew's encyclical letters, messages, conferences, travels around the globe, prizes and honours, and bilateral declarations with religious leaders of the world to increase environmental-protection awareness and care for the good creation of God have earned him the honorific title of "Green Patriarch."

The Ecumenical Patriarchate and the time for creation

On 1 September 1989, the Ecclesiastical New Year in the Orthodox Church, late Ecumenical Patriarch Dimitrios sent a message to the entire Christian world urging us to offer every year on that day prayers and supplications to the creator and maker of all, both as thanksgiving for the great gift of creation, and as petitions for its protection and salvation. He also requested all who are entrusted with the responsibility of governing the world's nations to act without delay to take all necessary measures for the protection and preservation of creation. It should be the day "when prayers and supplications should be offered for all Creation and for the reparation of the impairment caused to the natural environment."[6]

Today, 1 September is known as "Day for the Protection of All Creation" and has become established and adopted in full agreement with all the Orthodox Churches. It gave birth to the "Time for Creation" in the church calendar, starting on 1 September and ending on 4 October of every year, on the day when the Roman Catholic Church commemorates St Francis of Assisi, their patron saint of ecology. Among others, it was adopted by the WCC, by the Conference of European Churches (CEC-KEK), and by the Council of European Bishops' Conferences (CCEE) during the third European Ecumenical Assembly (EE3A) which met in Sibiu, Romania, in 2007.

Among some of the key concepts expressed by Ecumenical Patriarch Bartholomew over these many years are the following: ". . . Humanity has failed in what was its noble vocation: to participate in God's creative action in the world. It has succumbed to a theory of development that values production over human dignity and wealth over human integrity. . . . This is why nature 'groans and travails' in all its parts . . . [Rom. 8:22]"[7]

In November 1997 Ecumenical Patriarch Bartholomew declared that all aggressions to God's good creation are a sin: ". . . to commit a crime against the natural world, is a sin. For humans to cause species to become extinct and to destroy the biological diversity of God's Creation; for humans to degrade the integrity of the Earth by causing changes in its climate, by stripping the Earth of its natural forests, or destroying its wetlands; for humans to injure other humans with disease; for humans to contaminate the Earth's waters, its land, its air, and its life with poisonous substances: these are sins."[8]

Water as source of life

In a special message to the WCC, Ecumenical Patriarch Bartholomew addressed the issue of the vital importance of water, declaring: "Just as water is the essence of all life, water is also the primary element in the life of a Christian, where the sacrament of Baptism marks the sacred source of the spiritual life. . . . Water, then, signifies the depth of life and the calling to cosmic transfiguration. It can never be regarded or treated as private property or become the means and end of individual interest. Indifference towards the vitality of water constitutes both a blasphemy to God the Creator and a crime against humanity."[9]

Climate change: A matter of social and economic justice

In a letter to the WCC Working Group on Climate Change, and based on the fact that climate change is an unprecedented threat to the integrity and diversity of life on planet earth, Ecumenical Patriarch Bartholomew has denounced that ecological abuse as a sin against God and has recognized it also as a profoundly moral and spiritual issue:

> Climate change constitutes a matter of social and economic justice. For those who will most directly and severely be affected by climate change will be the poorer and most vulnerable nations—what Christian Scriptures refer to as our 'neighbours'—as well as the younger and future generations, the world of our children, and of our children's children. . . . There is a close link between the economy of the poor and the warming of our planet. . . . We must serve our neighbour and preserve our world with both humility and generosity, in a perspective of frugality and solidarity alike.[10]

In view of the ongoing worldwide financial and economic crisis, on 1 September 2010 he expressed: "It is important to note that the current grievous financial crisis may spark the much-reported and absolutely essential shift to environmentally viable development; i.e., to a standard of economic and social policy whose priority will be the environment and not unbridled financial gain . . ."[11] In this respect, the resources within the human spirit must be discovered in order to develop a sense of urgency and action.

Environmental seminars—international symposia on world rivers and seas

From 1994 to 1998, Ecumenical Patriarch Bartholomew organized five "International Environment Summer Seminars" at the Theological School of Halki at the Monastery of the Holy Trinity on the island of Heybeliada (Halki, in Greek), off the coast of Istanbul. The seminars were centred on the following key issues: religious education (1994); ethics (1995); communications (1996); justice (1997); and poverty (1998).

In June 2012 the Ecumenical Patriarchate and the Southern New Hampshire University convened as co-sponsors of the "Halki Summit 2012." The summit, envisaged as a conversation on environment, ethics, and innovation, stressed the importance of all participants having to pass along to the next generation a clean world with clean water and clean air, a pledge that intensely mobilized the grassroots "350.org Campaign."

In his keynote address, Ecumenical Patriarch Bartholomew stressed that the values of solidarity must prevail over greed:

> . . . Each of us is called to draw a distinction between *what we want* and *what we need*, or—more importantly—what the *world* needs. Greed and gratification reduce the world to a survival of the fittest; whereas generosity and gratitude transform the word into a community of sharing. . . . we know what needs to be done and we know how it must be done. . . . It is a long journey from the head to the heart . . . and an even longer journey from the heart to the hand. . . . There is no way of endlessly manipulating our environment that comes without cost or consequence. . . . this is a movement as critically urgent and as morally imperative as any campaign for *fundamental human and civil rights*.

The Ecumenical Patriarchate convened eight "Scientific International Symposia on World Seas and Rivers" between 1995 to 2009, bringing together scientists, environmentalists, policy makers, and religious leaders, drawing world attention to the degradation of the following bodies of water: (1) the Aegean Sea, 1995; (2) the Black Sea, 1997; (3) the Danube River, 1999; (4) the Adriatic Sea, 2002; (5) the Baltic Sea, 2003; (6) the Amazon River, 2006; (7) the Arctic Sea, 2007; and (8) the Mississippi River, 2009.

The "Amazon River Symposium," for instance, stressed the fact that the destruction of the Amazon rainforests would disrupt the whole global circulation of air and water, and thus the climate of the world. During "The Arctic Sea Symposium" it was affirmed that time is short and that actions must be taken together by all nations and act now, based on the frightening speed the Arctic ice is melting, plus the melting of Greenland's ice mantle. The consequences of both phenomena for Greenland and the rest of the world would be devastating. Seven years since the Arctic Sea Symposium, the ice mantles of the Antarctic continent are undergoing a similar phenomenon, while the needed actions have yet to be responsibly taken.

Interfaith Dialogue

Over the past decades, the WCC has established contacts with heads and representatives of Muslim, Jewish, Buddhist, Hindu, and Shinto religious leaders, among the faith communities. All religions agree that human beings have to be responsible stewards of the resources of our planet. These resources, as given to all generations by the divine, should be used for the benefit of every person without any type of distinction or discrimination.

Interreligious dialogue should be maintained by religious leaders and experts, and be enriched by the contributions of ordinary believers. We need to help one another and all well-intentioned people in building a better future for the entire human family. People of all faith traditions praise the divine, for they seek to understand their relationship to the cosmos. I want to share with you a few concepts on what is expressed in Judaism, Buddhism, Islam, and Hinduism.

Every year, at the New Year festival of Rosh Hashanah, Jews give thanks to God for the creation of the world because, although humanity has the role of steward, the Tanakh (the collected 24 books of the Hebrew Bible) shows that the earth is still God's possession: "The earth is the LORD's, and everything in it, the world, and all who live in it" (Ps. 24:1).

Moreover, the Torah (the Law, the first books of the Hebrew Bible) has numerous laws which attempt to redress the power and economic imbalances in human society and creation. There is a whole program in the Torah for creating a balanced distribution of resources across society (Ex. 22:24-26; Lev. 25:36-37; Deut. 23:20-21; 24:6, 10-13, 17). This is an expression of the concept of *tzadik*, which means righteousness, justice, and equity. It is the value which tries to correct the imbalances that humans create in society and in the natural world.

Other religions have also expressed these concerns. His Holiness the Dalai Lama says: ". . . our planet is our house, and we must keep it in order and take care of it if we are genuinely concerned about happiness for ourselves, our children, our friends, and other sentient beings who share this great house with us. If we think of the planet as our house or as 'our mother'—Mother Earth—we automatically feel concern for our environment. Today we understand that the future of humanity very much depends on our planet, and that the future of the planet very much depends on humanity."[12]

Buddhism also expresses that not only humankind but all beings have a common and equally precious existence:

> Buddhists are encouraged to love all living beings and not to restrict their love only to human beings. They should practise loving kindness towards every living being. The Buddha's advice is that is not right for us to take away the life of any living being since every living being has a right to exist. . . . Every religion advises us to love our fellow humans. Some even teach us to love them more if they belong to the same religion. But Buddhism is supreme in that it teaches us to show equal care and compassion for each and every creature in the universe. The destruction of any creature represents a disturbance of the Universal Order. . . . Since every creature contributes something for the maintenance of the planet and atmosphere, destroying them is not the solution to overcome our disturbances. We should take other measures to maintain the balance of nature . . .[13]

The environmental ethics of Islam is based on the special interest Prophet Muhammad had in the human and natural environment. His interest led him to teach: "Whoever plants a tree and diligently looks after it until it matures and bears fruit, is rewarded."[14] Central concepts for Islam are *Tawhid*, *Khalifa*, and *Mizan*.[15] They are also the pillars of its environmental ethics, constituting the basic values found in the Qur'an. *Tawhid* teaches the uniqueness and oneness of the Creator. As the Creator's *Khalifa* on earth, the human being is the steward and guardian of the creation, and thus is obliged to protect the environment and not to destroy it. *Mizan* is the principle of balance on which all creation works and maintains itself in a stable condition. Preserving the balance and stability of both marine and terrestrial ecosystems helps to safeguard the environment. Humanity should revere and nurture forces of nature, as nature is providing both material and spiritual sustenance to human beings.

Not only in the Vedas, but in later scriptures such as the Upanishads, the Puranas, and subsequent texts, the Hindu viewpoint on nature has been clearly enunciated. It is permeated by a reverence for life and an awareness that the great forces of nature—the earth, the sky, the air, the water and fire—as well as various orders of life, including plants and trees, forests and animals, are all bound to each other within the great rhythms of nature. The divine is not exterior to creation but expresses itself through natural phenomena. Let us recall the ancient Hindu dictum: "The earth is our mother and we are all her children."[16]

Faith-Based Organizations Network

As we have observed, people of faith share similar perspectives toward the care for human beings, for nature, for the earth. It is our task to succeed in developing a *common* vision and develop stronger bonds among the world's different faith communities. In this respect, during the 19th COP, an interfaith group of delegates met to continue discussing the formation of an interfaith group, in order to have a more constant and structured network of faith-based organizations (FBOs) attending the yearly UN conferences on climate change and advocating locally in their home countries and their governments for a more efficient result of the global international campaign to cope with climate change.

The interfaith group met several times during the two weeks of sessions. This movement is certainly inspired by the interest of the UNFCCC Secretariat—and, of course, of the different NGOs and IGOs—to have a first step completed during COP20-CMP10 in Lima, Peru, in 2014, in order to finally achieve a fair, just, ambitious, and legally binding agreement at COP21-CMP11 in Paris, France, in 2015. As a follow-up of these meetings and discussions both with the UNFCCC Secretariat and among FBOs, an Interfaith Liaison Committee was created to serve as a link between FBOs and the UNFCCC Secretariat.

Conclusion

In the book of Deuteronomy (16:20) we read: "Justice, and only justice, you shall follow, that you may live and inherit the land that the Lord your God is giving you." Justice and peace are needed: we must start our pilgrimage to justice and peace,[17] not remaining unmoved to the pain, to the loneliness, to the injustice, and to the effects and consequences of the war against nature. If we desperately wish to attain peace on earth, our home, we must be at peace with the earth, with creation. Peace on earth is reached with justice and love. We read in the in the writings of Saint Silvanus the Athonite (Russian Orthodox monk, Monastic Republic of Mount Athos, 1866–1938): "The heart that has learnt to love has pity for all Creation." With peace human beings are healed, respected, dignified, and honoured. Humankind adores its Creator and cares for human beings, for animals and plants, and for all the earth's natural resources that feed and sustain life on the planet.

The search for sustainable communities living with dignity, with justice, and with peace will need great personal sacrifice, while the current systems of greed and ambition will have to be replaced by systems of love and compassion.

Life is the most precious gift given by God, the Pantokrator, and the Maker of all things visible and invisible. Life is beautiful, and the beauty of the natural world cannot be compared to anything created by the human genius. But what is human life if it cannot be lived in full dignity, a dignity that can only be achieved in plenitude in a world filled with justice and peace? What would our future be like if there were no pure fresh waters to drink, clean air to breathe, and green forests and blue oceans to absorb and purify the atmosphere from the polluting carbon dioxide humanity is generating?

Through God's only begotten Son Jesus Christ, our God and Saviour, we shall attain the fully desired peace—peace with the earth, peace with creation—thus restoring the perfect balance, the harmony, and the beauty among us and the whole of God's good creation.

Despite the apparently negative signs of the times, we are still at a point where humankind is able to make a decision that could save the good creation of God from the otherwise inevitable desecration caused by human activities. The year 2014 gave us the opportunity to walk toward a fairer and sustainable future. The UN Secretary General Ban Ki-moon called for the "UN Climate Summit 2014" in New York on 23 September.[18] A remarkable number of world political leaders, heads of state and government, as well as leaders from business, finance, and civil society, attended the summit. They were challenged by Mr Ban Ki-moon "to crystallize a global vision for low-carbon economic growth and to advance climate action on five fronts: cutting emissions; mobilizing money and markets; pricing carbon; strengthening resilience; and mobilizing new coalitions."[19]

Once more, world religious leaders accompanied this transcendental meeting. The WCC, together with the World Conference of Religions for Peace (RfP), held the "Interfaith Summit on Climate Change," 21–22 September 2014[20]—immediately before the "UN Climate Summit." Religious leaders representing nine religions and 21 countries gathered in order to highlight the specific contributions that faith traditions bring to the international climate debate.

His All Holiness Ecumenical Patriarch Bartholomew expressed in an official message to the Interfaith Summit on Climate Change:

> Each believer and each leader, each field and each discipline, each institution and each individual must be *touched by the call to change our greedy ways and destructive habits.* For the message here is quite clear: unless we change the way we live, then we cannot hope to avoid ecological damage. This means that—instead of solely depending on governments and experts for answers—*each of us must become accountable for our slightest gesture and act in order to reverse the path that we are on*, which will of course also include prevailing upon governments and leaders for the creation and application of collective policy and practice.[21]

During the Interfaith Summit an inspiring statement[22] was signed, showing the togetherness and fellowship of the faith and religious traditions represented at the summit.

Let us pray to God, for whom everything is possible, to allow us to finally reach the needed shift of paradigm in our behaviour, habits, and desires, first in our hearts and minds followed by a profound change in our attitudes and actions, in order to enable all human beings of the present and future generations, without distinction, to enjoy the bounties of the good creation of God.

May the grace, the justice, the peace, and the love of Jesus Christ, our God and Saviour, be with us today, tomorrow, and always.

May God grant us the wisdom to continue along this path!

CHAPTER 9

Climate Justice and Environmental Protection

A Challenge to Churches in Asia, Africa, and Europe

Jochen Motte

The United Evangelical Mission (UEM) is an international communion of churches, with 36 members in Africa, Asia, and Germany. The UEM member churches and one diaconic member institution are located in DR Congo, Tanzania, Botswana, Namibia, Cameroon, Rwanda, Indonesia, Hong Kong, the Philippines, Sri Lanka, and Germany. When the international UEM was founded in 1996, its members committed themselves in its constitution to strive for justice, peace, and the integrity of creation. The UEM General Assembly adopted a "Statement on Corporate Identity" in 2008 that states, in part: "We believe that human beings are created in the image of God and therefore have inalienable dignity and rights. Therefore we promote and defend human rights. . . . we strive for the protection of the environment." At the same general assembly, the churches of the UEM discussed the issue of climate change and decided to give climate justice and environmental protection special attention and to establish this issue as a programmatic priority within the UEM. In the statement and decision, the delegates expressed the urgent need for action: "In Genesis 8:22, God promised His people that 'as long as the earth endures, seedtime and harvest, cold and heat, summer and winter, day and night will never cease.'"

The delegates, themselves beneficiaries of this promise, remain concerned about the effects of climate change and the increase in average temperatures that has already become a reality in certain regions. This development, a product of the human race, will have a major impact on the lives of billions of people and represents a challenge to the promise given by God, who is Lord of all creation.

While industrialized countries are most at fault for global warming, a result of carbon-dioxide emissions, the most life-threatening consequences will hit the poorest of the poor in the less developed countries of the global South. If these groups are no longer able to trust in seedtime and harvest, if cyclones and hurricanes destroy their livelihoods, if their access to clean water is cut off, if their environment is destroyed, they risk losing their human rights: the conditions necessary for economic survival could disappear for billions of people.

As stewards of God's creation, Christians must act immediately to stop global warming and to pressure the governments of the countries with the most responsibility for climate change to take effective and speedy measures to prevent further adverse consequences from rising temperatures.

As members of the UEM, we also see our own responsibility to protect and care for God's creation and to promote good stewardship in our regions and countries. This is why we have started and continue to promote the use of renewable-energy sources and sustainable agriculture and forestry to protect the environment and combat extreme poverty. The immediate action needed to stop global warming has made "climate and environment" a special priority for the UEM for the next four years. The UEM has been asked to support member churches in developing and implementing programmes and projects that address global warming and climate change in their own context and regions. The UEM has thus made the following decisions:

- Climate change and global warming will become a special programmatic priority for the UEM during the next four years.

- The UEM is tasked with developing adequate methods and providing necessary sources to support churches in addressing global warming and climate change through information sharing, awareness building, project support, and programme-related staff exchanges.

Since 2008, the UEM has broadened its approach to address climate justice and environmental protection and has developed activities, programmes, and initiatives in the following areas:

- Awareness building
- Advocacy
- Climate consulting
- Project support
- Climate/environmental responsibility in the UEM's own operations and policies
- Networking
- Climate justice and human rights
- Theology and spirituality

Building Awareness of Climate Justice in Germany, Africa, and Asia

In 2008 the UEM launched an exhibition for congregations and churches in Germany, called "Climate of Justice," that demonstrated the impact of climate change on vulnerable people in Africa and Asia. The exhibition aimed to raise awareness among people in Germany that the CO_2 emissions caused by Western lifestyles and industry affect people who have not contributed at all to these emissions but who still have to face the bitter consequences of increasing droughts and flooding—whose food security and basic social and economic human rights are in danger.

Awareness building with the same objective has also been done through publications, seminars, films, and other materials and actions, such as the annual International Youth Climate Action Day, when youth initiatives in different parts of the world are interconnected through Facebook, Skype, and the Internet. A special highlight of 2012 was the event at the Protestant School of College Evangelique de New Bell in Cameroon, where many children and young people took advantage of the opportunity to learn more about solar energy. The radio programme following this event reached thousands of listeners. The event was complemented by a concert and a writing workshop in which 300 children and youths from 16 different schools participated.

Advocacy on Climate Justice and Environmental Protection

Advocacy work with Western governments and in European societies

Since 2008, the UEM has joined or taken part in the establishment of several networks and platforms, such as the Climate Platform of Churches and the Climate Alliance of Churches and other civil-society groups, in order to lobby Germany and the European Union to make significant and binding commitments to decrease their CO_2 emissions. These commitments include limiting the rise of temperatures to no more than two degrees Celsius and taking strong responsibility to assist countries in the global South with their efforts toward adaptation and mitigation.

In view of the upcoming Conference of Parties (COP) in Paris in December 2015, and based on the decision by the general assembly of the World Council of Churches (WCC) to start a Pilgrimage for Justice and Peace, the German Platform of Churches on Climate Justice has initiated the planning process for a pilgrimage that will start in northern Germany and move toward Paris. The pilgrimage is to include the area in western Germany where various UEM member churches and the UEM Central Office are located. Pilgrims will be hosted by congregations and church people on their way and will take part in worship services and meditations. Local actions will take place to raise awareness. One stopping point for pilgrims will be the UEM in Wuppertal, the home of the UEM Central Office and the international conference centre called On the Holy Mountain.

Advocacy for marginalized people

In its advocacy work, the UEM strongly supports churches that are allies of those who have become victims of environmental destruction and the exploitation of natural resources. One such church is in North Sumatra, where indigenous farmers risk losing their forests because of operations by the Toba Pulp Lestari factory.

The pulp and paper company originally known as PT Inti Indorayon Utama started operations in North Sumatra in 1989, and people in the region have been protesting against the pollution from the mill ever since. Although it was closed by the government in 1998, the company reopened some years later under the new name of Toba Pulp Lestari. In 2009 the company announced the expansion of its pulp and paper production from 165,000 tons to 300,000 tons per year. This meant that new forests had to be cleared and prepared for industrial tree plantations. Some of the affected rainforests are located in Sipituhuta and Pandumaan, where indigenous communities have farmed resin from benzoin trees in an area of about 4,100 hectares for more than 300 years.

In 2013, 31 villagers who had been protesting against the company were arrested. NGOs and church leaders called upon the government to release the prisoners and expressed their solidarity with the affected communities. The UEM supported these churches in their advocacy work and the legal assistance they provided to the farmers who had been detained. Although all of them have since been released, the general threat to people and their livelihoods remains, a result of companies that profit from exploiting natural resources. The UEM international ecumenical team expressed its solidarity with the people in Pandumaan and Sipituhuta through a visit in 2012 (see below). The team gave recommendations and discussed these with church leaders in the region, who later on were able to work together to assist the communities when the arrests took place. Staff members of the UEM and church leaders visited the farmers in prison during their detention.

Addressing the negative effects of palm-oil production on people and the environment

Together with Bread for the World, the church-development agency of the Protestant churches, the UEM has published studies and fact sheets on palm oil and the Jatropha plant that indicate the risks of the increasing global demand (including in Europe) for food and biofuels. Such demand threatens forests, nature, and people in countries such as Indonesia.

For many years, Bread for the World and the UEM have partnered to call attention to the destruction of precious ecosystems and the violent social conflicts caused by the rapidly expanding cultivation of oil palms. Together these two organizations have taken a stand to ensure that the rights of farmers and indigenous people are being respected and protected. Most of those affected cannot expect any help from the state. In many countries the situation is influenced by conflicting laws, overlapping and uncertain common law, illegal operations, corruption, and a lack of legal certainty. Human-rights obligations are often insufficiently enforced or end up having no effect on the ground.

Bread for the World and the UEM have produced research based on these findings that asks whether voluntary agreements on sustainability between producers, retailers, and palm-oil processors, such as the Roundtable on Sustainable Palm Oil (RSPO), will be effective at addressing the aforementioned problems. Their study shows that there are severe shortcomings in the present RSPO system. Without significant improvement in elements such as monitoring and complaints systems and sanctions, standards like the RSPO will not ensure sustainability or protect human rights. Although the development of international agreements on sustainability between private producers and national as well as international companies is better than nothing, it is obvious that such kinds of agreements can only become effective within a national framework where the rule of law is enforced and governments fully comply with international human-rights standards.

Climate Advising for Churches in Africa and Asia

When climate justice and environmental protection became a priority in the work of the UEM in 2008, churches in Africa and Asia immediately expressed a need for further information, networking, and advice on the issue. The UEM therefore established two posts in 2010 that would provide climate consulting for Africa and Asia. Two experts have been appointed, one from each of the regions, and have started advising churches on climate and environmental concerns, project planning, monitoring and implementation, networking, and awareness building. The consultants have implemented a series of regional programmes in Asia and Africa in order to strengthen the capacity of churches to address the issue of climate change and environmental protection through their own project planning, theological reflections, and networking. Both consultants have also conducted programmes with congregations and churches in Germany in order to share their experiences from the global South.

Experiences from the work of the climate consultants have shown that churches are often willing and interested in starting and implementing projects and initiatives to protect the environment and the climate. Often there is a need for churches to improve planning, monitoring, and evaluation in order to approach foreign church agencies for funding. The climate consultants have contributed in a myriad of ways to empowering and professionalizing churches to manage complex communication and project applications with church-related agencies abroad. The consultants have also encouraged the churches to strengthen their cooperation and networking with NGOs and civil society in their respective local, national, and regional contexts so as to make use of available resources, know-how, information, and to avoid the unnecessary duplication of work and even counterproductive competition.

Project Support on Environmental Protection and Sustainable Development

The work of the climate consultants in Africa and Asia has encouraged churches to initiate and implement concrete projects in the areas of climate and environmental protection. Between 2012 and 2014, the UEM provided financial support to more than 30 such projects, including provision of stoves to enable people to reduce consumption of firewood; reforestation projects; alternative-energy projects such as solar lamps, biogas, and micro-hydropower plants; and initiatives on alternative farming and biofertilizers. In recent years it has become clear that the climate issue cannot really be separated from the environmental concern. In many places, heavy rainfall has induced flash floods, washing away houses and even killing people. One cause is global climate change, but other factors can contribute and worsen the devastating damage from these floods, including soil erosion from local deforestation and the construction of buildings on former farmland or forest areas that prevent water from draining off properly.

Climate/Environmental Responsibility in UEM Operations and Policies

In recent years the UEM has tried to adapt and change its administrative policies to take climate and environmental criteria into account in its operations and management. The UEM has included such calculations in all business travel for the last four years, including trips taken by the members of its governing bodies and all participants in the UEM international consultations, seminars, and workshops. Fifty thousand euros have been spent every year since 2013 to offset the CO_2 emissions. This compensation is done through different projects implemented by an institution called Climate Collection. In the meantime, the UEM has started to initiate its own Clean Development Mechanism (CDM) project within the Anglican Church in Rwanda. Local communities will receive carbon-friendly stoves, which will reduce the consumption of firewood and CO_2 emissions by up to 70 percent. Furthermore, the UEM has begun adapting its operations in the field of real estate and transport, for example by purchasing cars with low fuel consumption or purely electric engines and renovating the UEM-owned buildings by installing new heating technologies, insulation, and solar panels.

Ecumenical Learning and Networking in Cooperation with the WCC

Sharing experiences of ecumenical church partnerships

It has already been mentioned that the UEM, in the process of addressing climate justice and environmental protection, has become a member of several networks. As a communion of churches, the UEM is an associated member of the WCC. This provided an opportunity five years ago for the UEM to become a member of the WCC Working Group on Climate Change. A UEM team was actively involved in the International Peace Convocation of the WCC in Jamaica that took place from 17 to 26 May 2011. Supported by representatives from its member churches, the UEM conducted two workshops at the convocation, one on impunity and one on climate justice. Together with the Protestant Churches in Germany (EKD), the UEM produced a film for the convocation that documented how UEM churches in Germany, Sumatra, Java, and Papua are striving to protect the integrity of creation by promoting rainforest protection, biogas production, and solar lamps.

One of the examples featured in the film reflected on the cooperative partnership of a church circuit of the Westphalian Church (EKvW) in Germany with a church circuit of the Protestant Church in West Papua/Indonesia (GKI): young people in Germany started a project to manufacture solar lamps, the costs of which were covered by donations from members of congregations in Germany. The lamps were then shipped to West Papua, giving people in the highlands living without electricity, including young students, the chance to light their houses in the evening. A workshop was later established in Jayapura, the capital of the province, where broken lamps could be taken for repair. On the one hand, this ecumenical cooperation enabled pupils in particular to continue studying after dark, while also saving on costs for petrol. But what is even more important is that young people in Germany could gain an understanding of what it means to live without electricity and could then reflect on the privileges of their own lifestyles. They were able to reach out to people far away in West Papua who might be experiencing the consequences of climate change, even though the Papuans themselves had hardly contributed to CO_2 emissions at all in their entire lives.

Conducting international ecumenical team visits

As a continuation of one of the core themes of the consultation, "Peace with the Earth," the UEM invited representatives of its member churches from Africa, Asia, and Germany to conduct team visits to UEM member churches in Papua and Sumatra from 1 to 13 May 2012. The teams had the opportunity to see the impact of mining, palm-oil production, urban pollution, and deforestation on the environment and the local communities, and could share and discuss these challenges with church leaders from the regions.

In Papua, for example, one of the ecumenical teams met with indigenous people in the area of Merauke. Nomadic communities living there for centuries had already sold their land to the Merauke Integrated Food and Energy Estate in West Papua (MIFE), a company that was founded to serve the demand for land of domestic agribusiness conglomerates. There are plans to convert 1.2 million hectares into mega-plantations. In response to pressure from a branch of MIFE that intended to produce sugar cane, Papuans leased the company thousands of hectares of land, including the resting places of their ancestors, at a fraction of the current market rate. The deal divided communities, showing what an important role churches—which are present even in rural and remote areas—could play by warning people in time of the consequences of losing their land before it is too late and irreversible, legally binding decisions have been taken.

Church leaders and experts shared their experiences and discussed the results of the aforementioned international team visit at the general assembly of the WCC in Busan, South Korea, in November 2013 and at a *madang* workshop. The speakers urged the WCC to take up climate and eco-justice as a key priority for the coming years.

Starting a "Pilgrimage for Justice and Peace"

Half a year after Busan, in May 2014, the UEM hosted a meeting of the WCC Working Group on Climate Change at its conference centre in Wuppertal, Germany. In addition to attending the internal meeting of the group, the 30 participants started planning for a Pilgrimage for Justice and Peace by visiting the Evangelical Church of Westphalia and the Evangelical Church in the Rhineland. The visit gave them a chance to become acquainted with the climate justice programmes within these two big German Protestant churches, as well as to share their own experiences with representatives from the leadership of these churches. Participants also visited a huge soft coal open-pit mine in the Rhineland. The coal used for energy production contributes significantly to Germany's high CO_2 emissions. The group visited a congregation in the same area, where church members had renovated their church building to be carbon-neutral and use very little energy. All of the energy needed for electricity and heating is generated by the building and its solar-panel installations, while the insulation of external walls has reduced energy consumption drastically.

Climate Justice and Human Rights

On 10 December 2008 and 2011, the UN-declared Human Rights Day, the UEM launched campaigns addressing climate change and land grabbing as human-rights violations. The campaign motto in 2011 was a verse from Leviticus 25:19, "The land will give its fruit, and you will have all you want to eat and will live in safety." In recent years, the commodity of "land" has developed into an object of investment and speculation, highly sought after by states, international and national corporations, banks, and investors. The growth in the global population and the rising demand among industrialized and developing countries for food and biofuels, along with other raw materials, has led to a run on the last huge tracts of land that can still be bought or leased cheaply. As part of the campaign, the UEM informed churchgoers in Germany about the efforts of churches in West Papua, Indonesia, and Tanzania to raise awareness among the indigenous peoples and peasants who risk losing their land because they have been lured with false promises into giving their land away to investors, who then convert the forest into plantations.

From the beginning, the UEM has addressed climate justice and environmental protection as a human-rights issue, because it understands that climate justice is closely related to human rights. When people lose all of their food because of flooding or drought and may even have to leave their lands or lose their homes, basic human rights are being threatened.

The UEM published a study on climate and human rights in 2013 in collaboration with the WCC to try to call more attention to international human-rights standards and to mechanisms for protecting them. Churches were encouraged to make use of their presence, locally and globally, and to amplify the voices of victims of climate change from a human-rights perspective. This seems to be urgently needed, since the main international forum to address and negotiate climate protection—the Conference of Parties (COPs)—has not yet demonstrated any real progress. It is obvious that climate protection—at least the human-rights aspect—will not be achieved through mutual voluntary commitments by states to limit the consequences of climate change based on their mutual interests. States have an obligation to act to protect victims. This obligation of the state not only comprises actions toward its own citizens, but includes an extraterritorial dimension if one country's CO_2 emissions contribute to the negative impacts of climate change elsewhere. A human-rights-based approach should become much more of a priority in order to hold states accountable for immediate action. Such action should be monitored by adequate international human-rights mechanisms within a human-rights legal framework that must be developed.

Theology and Spirituality

Climate change and environmental destruction are serious challenges to church life rooted in biblical tradition. The integrity of creation has become a major concern for UEM member churches over the last six years. In 2009, the UEM started a process to collect songs, liturgies, and biblical reflections from its member churches, with their various cultural backgrounds in Africa, Asia, and Europe. This content was compiled into a booklet on climate and spirituality that was published in 2013 in five different languages, making these materials available for churches to use in worship and church life. Further materials have been collected and presented on the UEM homepage. This project could only be realized because the churches in the UEM had already begun to reflect on creation theology in their respective cultural contexts. Climate and nature have become a matter of biblical reflection, theological debate, spiritual life, prayer, and action.

In 2014, the UEM joined the project by Bread for the World and the WCC to conduct a global survey on Ecotheology, Climate Justice and Food Security (ETCF) in theological education and Christian leadership development. This project will highlight and make visible the contributions, resources, and perspectives of eco-theology, climate justice, and food security, especially from churches in the global South. Ecotheology, climate justice, and food security shall become an integral part of theological education and ecumenical formation at all levels and international exchange and networking on this issue shall be enhanced and promoted.

Outlook

It has been greatly encouraging to see how far churches in the UEM have come since 2008 in addressing the issue of climate justice and environmental protection in all its dimensions. In the coming years the UEM will continue to strengthen its members in its efforts to bring these concerns to the grassroots and to integrate them into the spiritual life and diaconic action of its congregations. At the same time, the UEM will support churches that are continuing or even have already started on advocacy work to put a stop to land grabbing and the exploitation of natural resources in places where biodiversity and people's livelihoods are endangered.

It is also obvious that what has been achieved since 2008 would not have been possible if implemented in an exclusive way. Cooperation, sharing, and networking with other ecumenical bodies and communities, especially with the WCC, has been key to the success of this process.

The member churches in the UEM are glad to join the Pilgrimage for Justice and Peace of the WCC and to be part of the global church action and advocacy. Governments must be urged to make binding global commitments to reduce CO_2 emissions, and soon, so that climate change can be stopped before it is too late.

CHAPTER 10

From Climate Justice to the *Energiewende*

The Debate in Germany and Churches as Change Agents

Richard Brand, Katja Breyer, Klaus Breyer, and Eva-Maria Reinwald

The Debate on Climate Change within Germany[1]

Climate change and its global impacts have become integral parts of the inner-German political discourse these days. There is a cross-party political consensus that combating climate change is one of the major political challenges of the coming decades.

At the international level climate protection was recognized as a field of action at an early stage, one in which Germany had the possibility to distinguish itself in a pioneering role and gain an economical advantage from climate-protecting measures—for example, by the export of technologies. Although Germany unfortunately has lost this pioneering role in recent years, climate protection continues to be an important political objective. The measures taken, however, are not profound enough to match Germany's responsibility. Also, for reasons of global justice, church actors push for an ambitious implementation of climate protection.

Climate change exacerbates poverty and threatens life. Global warming destroys natural resources, undermines opportunities of development, and reinforces injustice. It poses tremendous challenges for the political world, the economy, society, governments, and the church; for many, it is their survival that is at stake, but for everyone, it is a question of a dignified existence. Climate change hits the poor in developing countries hardest, that is, those least responsible for it, and thus further deepens the gap between rich and poor. Industrialized countries like Germany are mainly responsible for climate change. Thus, reducing emissions in Germany and other industrialized countries is fundamental for climate justice.

Effective climate protection can only be achieved through a change of course in society. The German Advisory Council of Global Change (WBGU) names this necessary change "The Great Transformation."[2] Fundamental changes are necessary across all policy fields and levels—nationally and internationally, in business and in society, as well as in the private sphere. Not merely a technological revolution is required to move from the "fossil" age of apparently boundless growth to a low-carbon, environmental and socially acceptable global economy but, rather, a far-reaching cultural change, a rejection of development models based on growth, especially in the highly developed industrialized countries. One fundamental change would be a successful energy transformation (*Energiewende*) in Germany to realize a climate-friendly and low-risk energy supply, which would be a central contribution to climate justice.

The objective of the *Energiewende* is to replace the current energy system, which relies on coal, oil, nuclear power, and gas, with a new energy supply based on renewable energies, namely wind power, solar energy, hydropower, biomass, and geothermal energy. This can only be achieved when energy saving and energy efficiency have priority.

Germany has set ambitious targets for reducing emissions: greenhouse gas emissions are to be reduced by 40 percent by 2020 and by 80 to 95 percent by 2050, compared to the reference year 1990. Germany currently generates approximately 45 percent of its electric power from coal, causing great amounts of greenhouse-gas emissions. Existing and planned coal power plants are the main obstacles for achieving the reduction goals and for implementing the *Energiewende*. To overcome this and other obstacles, several actors of the civil society and also churches are demanding a federal climate-protection act to ensure the necessary measures for climate protection. But the government is not willing to adopt such an act. In 2013 two German federal states, North Rhine-Westphalia and Baden-Wuerttemberg, adopted their own climate-protection acts. North Rhine-Westphalia is especially important, because 30 percent of the energy production in Germany is generated in this federal state, mainly from coal power plants.

Currently in Germany a strong confrontation is taking place between the coal-fired power stations and the booming renewable energies, between the centralized and decentralized, and between the big energy companies and small energy suppliers. This is due mainly to the Renewable Energy Sources Act (EEG), which stipulated that by 2014 approximately 25 percent of the electricity produced in Germany come from renewable sources. It has created a massive surge in the increase of wind power, solar technology, and bioenergy and has become a model for more than 100 national or regional laws concerning renewable energies worldwide. The EEG also prompted more decentralized energy production, sometimes owned by citizens. Many renewable-energy cooperative societies have been established. Churches are also engaged in energy production by photovoltaic systems and windmills.

Furthermore, the nuclear-power phaseout is a central part of Germany's energy transition. The disasters at Chernobyl and Fukushima illustrated clearly that nuclear energy is too risky. By around 2022, the last nuclear plant in Germany is to be shut down. Due to the insufficient emissions-trading system, however, the nuclear phaseout has led to an increasing use of coal for electricity production.

The European Emissions Trading System (EU-ETS) is the main instrument in the European Union (EU) to lower greenhouse-gas emissions in industry and the power sector. It puts a cap on annual emissions that is reduced over time and allows the trading of greenhouse-gas emission allowances. In the opinion of many civil-society actors and also of churches, the EU-ETS is lacking in ambition and has too many loopholes—an outcome that comes as no surprise, given that policymakers had to make concessions to strong electricity and industry lobbies to get the system launched at all. In its current state, EU-ETS is not effective for climate protection. Many NGOs and also churches are lobbying for a more ambitious ETS in Europe.

Another great challenge in the necessary transformation is the reduction of energy and resource consumption. Using renewable energies and energy efficiency must be complemented with a policy of sufficiency or self-restraint. Otherwise, the dynamics of expansion will drown out the success of greater efficiency and consistency. This means a profound debate over the question of "How much is enough?" and a deep economic and social change. Churches can and must play a crucial role here by advocating for sustainable patterns of economy and consumption and a lifestyle based on justice and sufficiency.

For this, it's also necessary that churches take credible steps toward a more sustainable society. The 2008 Synod of the Evangelical Church in Germany (EKD) set clear and ambitious goals when it recommended that its member churches reduce their carbon emissions by 25 percent (in comparison with emissions in 2005) by 2015. Many regional churches set up climate-protection concepts to achieve these goals and a many activities have been set in place. But much more will be needed to reduce further the emissions. This is a huge challenge for churches.

Responsible acting in their own ranks, political lobbying, and ecumenical cooperation are the action fields where the Evangelical churches in Germany advocate for the integrity of creation. The remainder of this essay will describe the theological basis of this commitment, point out the efforts of our churches concerning climate protection, and set perspectives of action against the background of experiences and impulses achieved in global ecumenism.[3]

The Engagement of the Church in Climate Justice and Caring for Creation

Theological insights for engaging in climate protection

Really long-lasting and sustainable economies and lifestyles demand a fundamental societal reorientation, a transformation, a change in awareness and mentality as well as in policies—in other words, "turning around to life" is needed. "Turning around" (*metanoia*) in the biblical sense means the overcoming of alienation from God. Turning around does not mean turning back to ostensibly better days of yore, but instead, biblically speaking, shifting to a radical new focus on God's commandments and affirmations of salvation. It refers to a renewal of our thinking and action through our faith in the gospel of Jesus Christ. "Do not be conformed to this world, but be transformed by the renewing of your minds, so that you may discern what is the will of God" (Rom. 12:2).

Turning around in the Christian ethical sense means, first and foremost, perceiving the challenges of the present as well as one's own entanglement, in order that one might leave life-threatening ways and strike out in a new direction at last. Turning around or, better said, this "turning back to the path of life" needs points of reference. The guideline of a "sustainable development, viable for the future" is one such important point of reference. Sustainability is no genuinely biblical term but, nevertheless, it is explicitly placed in the center of the Jewish-Christian faith in creation.

A more beautiful and exact description of sustainable development than that given in the second creation narrative is hardly possible. "And the LORD God took the man, and put him into the Garden of Eden to dress it and to keep it" (Gen. 2:15). "To dress and to keep!"—God gives humans the order to dress their world, to cultivate the soil, to farm the land, but without destroying the natural resources. Humans ought to learn to live and work in a way that preserves continuity of life on earth.

On the seventh day God crowns God's work of creation with rest, the sabbath. Humans are not the crown of creation! Only the calm finalizes creation. "God looked at everything he had made, and he was very pleased" (Gen. 1:31). The biblical tradition regards "God's rest" as a role model for humans and therefore, at the same time, as an important measurement for all acting. God's acting shows that salvation will not be found in eternal progress, in the permanent increase of "faster, higher, further." Creation, rather, is completed in our ability of "leaving alone." The regulations of the sabbath and the sabbatical (Ex. 23:10ff.; Lev. 25; Deut. 15:1ff.) explicitly include the nonhuman creation. In the sabbatical, all the land is given back to its original owner. That puts a stop to land speculation and accumulation of property. It reminds us that the land, the soil, does not belong to humans but that it is the "earth of the LORD" (Ps. 24:1). The guiding principle within the belief in creation is a plea against anthropocentric immoderation and for a careful, low-risk handling of nature: do not try to get the most out of it—neither of nature nor of humans!

If we as Christians want to know how we can live from the promise of "a new heaven and a new earth" (2 Pet. 3:13), the "dressing and keeping" mentioned in the story of paradise and the careful, responsible stewardship of the first creation story—in the Priestly Source—includes the model of sustainable development. We must accept limits and should not try to get the most from humans and nature. More than ever, the rights of current and following generations must be protected.

The freedom for which Christ set us free (Gal. 5:1) also means liberation from the—ultimately pointless—constraint of self-redemption at the expense of others. The freedom for which Christ set us free means the freedom for self-limitation and political and societal acting, for being there for others. Such a self-limitation and acting is oriented toward dignity and freedom of other people, coming generations, and the conservation of creation. The freedom for which we as Christians are set free means a claim and a request at the same time: for a "sustainable development"! Sustainability—and thus also climate protection—is constitutive for the Judeo-Christian belief in creation. It is a guiding principle of Christian social ethics with numerous ecumenical roots. If we as churches want to be committed to sustainable development, we find ourselves in the center of our faith and our societal and ecological responsibility.

Recognizing the need for change

There are 22 Lutheran, Reformed, and United regional churches that form at present the Evangelical Church in Germany (EKD). The EKD is a public-law corporation, as are its member churches. The EKD has the following elected governing bodies: the Synod, the Council, and the Church Conference. German Evangelical Church structures are based on federal principles at all levels. The EKD carries out various joint tasks that are delegated by its members.

The EKD as well as its member churches have a long tradition in participating in public debates about political, social, and economic issues. The topic of sustainable development has been discussed several times at the annual Synod. An intensive engagement for preserving creation and a sustainable energy supply started in 1986, triggered by the nuclear catastrophe of Chernobyl. Since then the churches of EKD regularly have been taking positions and actions on energy and climate issues—within their churches, but also within society and politics. In recent years climate justice has become more and more a focus of the EKD, due also to the reports from ecumenical partners on the consequences of climate change for their lives—for instance, for sustaining agriculture and food supply. In 2013, for example, the Synod of the EKD addressed policymakers in Germany with the declaration, "'There is enough for everyone'—World Food Security and Sustainable Agriculture," calling them to do everything in their power to secure nutrition for the world population and to support them in adaptation to climate change.[4] Churches and their members were requested to consistently shape the way in which they conduct economic activities and life patterns in order to meet the objective of sustaining global food security and preserving God's creation.

The declaration refers to the path-breaking memorandum, "'Turning to life'—Sustainable Development in the Context of Climate Change," that the Council of the EKD published in 2009.[5] "Churches can do what is necessary in that they themselves repent and return to a lifestyle that upholds the core values of justice and sustainability"[6] is one of the key messages in this statement.

The Council declared in the memorandum that they are aware that the churches, involved in the dominant economic structures, have a direct responsibility to seek clarity about what is at stake and how they can contribute to viable solutions. They perceive themselves in a role of being defenders of the vital interests of the people suffering from climate change. Looking at climate change, but also at the recent global financial and economic crisis, it is obvious that new economic and political approaches are necessary. The EKD Synod in 2008 already had indicated the need for new sustainable development models and intensive action for climate protection.

The memorandum is an appeal to respond to the several crises in a spirit of solidarity, support, and sustainability. The document examines the influence of climate change on development and spread of poverty and proposes a biblical and theological perspective in identifying ways of repentance in politics and society. The concept of development based mainly on economic growth is not considered as viable. Instead, some core elements of a sustainable development model are described in greater detail.

In 2012 the EKD and other church institutions co-organized, along with the umbrella association of the trade unions and a large environmental network, a national conference on the social and economic transformation of our societies toward sustainability. Together with politicians and other stakeholders they discussed necessary strategies and measures in politics, economy, and society—for example, to reduce greenhouse-gas emissions, to implement a sustainable-energy supply, to create good jobs, and to achieve a fair distribution of resources in society. Additionally, several workshops and conferences have been organized by Evangelical academies and other institutions on different levels and there will be more to come.

Since churches are convinced that the existing development model based on growth, practiced in industrialized countries, is not sustainable, they stress the importance of the so-called ethics of sufficiency or ethics of enough in these and other debates.

> An "ethics of sufficiency"/"ethics of enough" is not primarily an ethics of sacrifice, but an ethics of the "good life", since it also advocates liberation from waste and materialism. It leads us freely towards a very different ethics; one of the right measure. It is rooted in the abundance that we celebrate and experience in Christ. It is inspired by the Biblical hope for God's kingdom, in which all people are promised freedom, justice and reconciliation. With this promise in sight, we step out onto the path of repentance and change today. Then, there will be enough for everyone![7]

Acting credibly

When our church passes on the notion of integrity of creation and claims a complying sociopolitical acting, the question arises if the church itself complies with these requirements. Churches could and should be exemplars when it comes to taking responsibility, changing lifestyle and habits, and creating a new awareness. But congregations, church institutions, and individual believers are often still far from this: insufficiently insulated buildings spoil energy; information for congregations printed on fresh-fiber paper contributes to deforestation; individuals traveling to a church committee meeting prefer using a private car rather than public transport; and Christians do not take fewer vacations by air than do their non-Christian neighbors. Programs and projects concerning climate and environment protection which move in the right direction with regard to credible action and which encourage churches to further engagement are presented below. They have been realized in many Evangelical regional churches and Catholic dioceses.

The* Grüne Hahn*: An environmental management system for churches

The "*Grüne Hahn*" is an environmental certificate for congregations and church institutions. It states that certificated congregations or institutions have implemented a functioning environmental-audit scheme, which means that they continuously try to improve their environmental balance. The *Grüne Hahn* corresponds to the EU Eco-Audit EMAS (Eco Management and Audit Scheme), but is adapted to the church context. After a congregation does stock-taking in the fields of electrical energy, heating, water, buildings, procuring, and waste disposal, they agree upon measures, responsibilities, and time periods for improving and monitoring their environmental balance. More than 500 congregations and institutions have already introduced the system and achieved considerable improvements in the environmental sector. In the Evangelical Church of Westphalia about 130 parishes participate in the auditing. Their consumption of thermal and electrical energy has been reduced by about 13 percent compared to the baseline year. This reduces the CO_2 pollution of the atmosphere by thousands of tons and the congregations save considerable financial sums.

The *Grüne Hahn* is a growing movement in the ecclesiastical context. These activities within the congregations also have a positive impact on the ecological behavior in the private sphere of the church members. In order to facilitate the introduction to the protection of resources and climate, a simplified energy-management tool was added to the *Grüne Hahn*, which permits a reduction of energy costs within a narrow time frame and with low financial investments.[8]

Climate-protection plans of the regional churches

How can a regional church realize an adequate, long-term reduction of greenhouse gases? Climate-protection "concepts" of the regional churches try to answer to this question. Since 2008 the German Federal Ministry of the Environment has promoted the concept development and the introduction of climate-protecting measures of regional churches and dioceses. In these concepts the areas of properties and buildings, mobility, and procurement are evaluated measurably and in detail. Based on this evaluation, the aim of the Evangelical Church of Westphalia, for example, is the 40 percent reduction of CO_2 emissions by 2020 (baseline 1990). Up to now, nine of the 20 Evangelical regional churches have developed climate-protection projects and several others are currently in planning/development. Thereby the implementation of the strategy depends on the structures of the individual regional congregations. One task of the climate-protection agency of the Evangelical Church of Westphalia, for example, is the development of suitable mitigation measures, their announcement on all church levels, and the support of their implementation. As a first important step the climate-protection agency has developed an energy-management system for church districts and congregations and is actually introducing it in close cooperation with the building directorate of the regional church.

Climate-friendly regional synods

As church parliaments, regional synods are the highest executive organ of an Evangelical regional church. Ecclesiastical law and financial rules are decided and important theological and ecclesio-political issues are discussed there. In 2011 the Evangelical Church of Westphalia started to shape its synods in a climate-friendly way. The synod's impacts on the environment (e.g., caused by mobility and accommodation) were measured and improvements have been reached by taking appropriate measures (e.g., a reduction in paper consumption). The CO_2 emissions, which can't be avoided despite improvements in the organizational process, are balanced via the compensation fund "*Klima-Kollekte*" (see below).

Klima-Kollekte*: A compensation fund by Christian churches*

"*Klima-Kollekte*" is the CO_2 compensation fund organized by the Christian churches in Germany. The fund was launched in 2011. By investing in sustainable climate-protection projects in countries of the global South, the fund compensates unavoidable CO_2 emissions caused by organizations, communities, congregations, and individuals because of flights or events. From its customers *Klima-Kollekte* receives 23 euros per ton of CO_2 emissions. In 2013 the total amount received by customers equalled nearly 290,000 euros. This corresponds to 12,600 tons of CO_2 emissions. The number of customers that are using the compensation fund is increasing. However, it is still only a small percentage of the "church-based" CO_2 emissions that are compensated.

The compensation projects either promote the development of renewable energies or enhance energy efficiency and are always linked to the objective of poverty reduction. There are projects like biogas plants and solar lamps in India or energy-efficient cooking methods in South Africa. Partners of the Protestant and Catholic development agencies, Bread for the World-Protestant Development Service and MISEREOR, carry out these projects. All projects of *Klima-Kollekte* with an annual CO_2 reduction of more than 5,000 tons of CO_2 are Gold Standard (CDM) certified. The Gold Standard ensures that the projects meet the highest social and ecological standards.[9]

Eco-fair procurement

Sustainable consumption and eco-fair procurement are important action fields in which congregations and institutions can prove their credibility. Germany-wide, the ecumenical project "*Zukunft einkaufen*—Credible Economics in Churches" supports congregations, church administrations, and educational facilities with aligning themselves toward ecological and social criteria. Congregations get recommendations for the most diverse range of products, from recycled paper and environmentally friendly cleaning products to eco-electricity and fair-trade coffee.[10]

Increased attempts were also made to root ecological procurement in the social-welfare organizations Diakonie and Caritas. In Germany, Diakonie and Caritas are providers of numerous hospitals, nursing homes for the elderly, playschools, and advisory centers, and they exert a significant buying power. If these institutions increasingly focus on sustainability, enormous options open up for reducing their environmental impact. Besides procurement and use of energy and food, the focus is put on textiles and care, cleaning, and hygiene products. But the launch of ecological procurement at Diakonie and Caritas faces major challenges due to specific product and service requirements—for instance, in hospitals—as well as the competitive situation relative to other social-services providers.

Educational work

Numerous regional churches are committed to educational projects and campaigns that motivate and support youngsters and/or adults to stand up for climate protection and climate justice—in their personal environment, their congregation, and community. Some examples for these educational projects are:

- The information center "Climate Justice" of the Center for Mission and Ecumenism of the Evangelical Lutheran Church in Northern Germany, which sensitizes users especially to issues of climate justice with events, workshops for school classes, confirmation groups, competitions, and educational material.[11]
- The campaign "*Klimawandel—Lebenswandel*" of the Evangelical Church in Central Germany, which started in 2011 to encourage church members to reduce CO_2 emissions and offered materials for services and pedagogical work in churches.[12]
- The campaign "*Gutes Leben für alle*" of the Roman Catholic Diocese of Speyer, which has collected over 1,000 answers from church members to the question, "What contributions can all of us make to reach a sustainable lifestyle?"[13]

- The youth campaign "Mission: *Klima retten*! powered by heaven" of the Evangelical Church of Westphalia. Its travelling—and participatory—exhibition "EcoCity" is shown at different places. It encourages young people to scrutinize their lifestyle and to develop climate-friendly alternatives.[14]

Ethical investment

Churches also try to contribute to climate protection with ethical investment. The Evangelical Church in Germany, for instance, offers guidelines for ethical and sustainable investment for its member churches.[15] Following these guidelines, money investments should not only be made from an economic point of view but also in consideration of Christian values and thus be socially acceptable, ecologically fair, and sensitive toward intergenerational aspects. Countries that didn't sign the Kyoto Protocol are therefore excluded from monetary investments. Financial investments of churches should prefer enterprises that support the reduction of raw-material consumption, of water consmption, and of energy and pollutant emissions, as well as enterprises that refine and promote the use of renewable energies, that formulate their own environmental guidelines, or that have implemented their own environmental-management system. Regarding real estate investments, attention should be paid to whether they are constructed and operated in climate- and environmentally friendly ways.

Political advocacy for climate justice

The churches in Germany are influential political actors. They participate in the public debate on a wide range of topics, not only those topics that are related to religious issues but also on all questions that are subsumed under the category of global responsibility. Fighting for climate justice and the integration of development and social aspects into the climate-change discussion is one essential field of political intervention that churches and faith-based organizations are trying to influence. The levels of intervention range from a global perspective (international climate-change conferences) to local municipalities. A reasonable division of work among the different institutional church-related actors has been established. This includes regular exchanges to assess climate and environmental politics, to share activities, and to develop strategies. From the beginning of climate-related political advocacy a close cooperation with Catholic institutions and actors has taken place.

The global level

Bread for the World—Protestant Development Service[16] is the globally active development and relief agency of the Evangelical churches in Germany. Key issues of the agency's work are food security, promotion of health and education, access to water, strengthening of democracy, respecting human rights, keeping peace, and the integrity of creation. In regard to the international climate negotiations, Bread for the World is the pivotal actor of the Evangelical churches. They have professional staff working on issues like mitigation, adaptation, climate finance, and low-carbon development. Together with other civil-society organizations and their project partners from all over the world they are monitoring the climate negotiations, lobbying the national negotiators, and organizing side-events and campaigns.

The EU level

The European Union is playing an important role in regard to climate politics, because there is a joint representation of the EU member states at the UN Climate Change Conferences. Therefore, the EU is one of the main lobbying contacts for civil-society organizations. Bread for the World is a member of APRODEV (now ACT Alliance EU),[17] the association of World Council of Churches (WCC)-related development organizations in Europe. Its main objective is to influence decision-making processes in the European Union institutions as these affect developing countries, in order to promote justice and peace and the eradication of poverty. APRODEV pursues rights-based development from a faith-based perspective.

The overall objective of the APRODEV Working Group on Climate Change and Development is global climate equity. The group is strategically influencing European Union policies with regard to climate change, aiming to ensure that they safeguard and are coherent with the EU's stated poverty-eradication and development policies, and that the EU recognizes both its historic responsibility for climate change and to protect the rights of developing countries to do such development.

The national and federal state level

Compared to other industrial countries, Germany is among the most active nations in regard to climate protection and opted for its own ambitious reduction targets several years ago. However, in recent years the German international climate policy has lost some momentum. Nevertheless, environmental protection and climate policy are top-ranking issues that have gained a high level of consensus in politics and society during the last few decades. The ecological movement has a long tradition, starting in the 1970s with the protest movement against nuclear energy. Many Christians, parishes, and regional churches were and still are active in the environmental movement. Without this broad and strong ecological awareness in German society, the two major political decisions to phase out nuclear energy and to start the energy transition (*Energiewende*) wouldn't have been possible.

Already in 1987, the synod of the EKD decided that nuclear energy is not compatible with the biblical mandate to respect the integrity of creation. They recommended that the energy system should be transformed as soon as possible toward renewable energy sources. This position was confirmed by the following synods in 1998, 2006, and 2008. Innumerable decisions on various ecological aspects have been taken by the EKD and by all the regional churches. The majority of the regional churches have full-time environmental officers to strengthen eco-management and to raise awareness inside the church and the parishes. Through the EKD working group of environmental officers (AGU), mutual support and a high coherence are maintained and joint lobbying activities are promoted.

As the governments of the federated states in Germany also execute energy policy, there are additional opportunities for the regional churches to influence decision making. North Rhine-Westphalia (NRW), the biggest federated state in Germany, produces nearly 40 percent of the CO_2 emissions in Germany due to the large fossil power plants using lignite. The large lignite open-cast pit mining areas are located in the area of the Evangelical Church in the Rhineland. Many districts are severely affected by the mining. Some municipalities and parishes had to move and were resettled. Therefore, many local churches have a long tradition of resistance against the open-cast pit mining. Based on these experiences, programs to promote renewable energies, to increase energy efficiency, and to practise eco-management have flourished. The regional churches have established close relationship to the governments and its ministries, promoting dialogue about climate and energy issues. Members of the Evangelical Church of Westphalia were highly involved in the process to adopt a climate-protection law in NRW and to design a comprehensive climate-action plan.

To put it in a nutshell: political advocacy to promote environmental issues has an important role in German Evangelical churches' activities for the integrity of creation. Long-term commitments, full-time staff, supportive network structure, professional competence, openness for dialogue, and public visibility are some factors supporting Evangelical churches in Germany being respected stakeholders in the environmental and climate debate.

Acting in the Ecumenical Context

Climate protection is a task that challenges churches worldwide. At the same time, the international relations within the global ecumenism can be especially useful for conceiving the impact of climate change, boosting climate protection, and learning from each other. Below is a selection of ecumenical projects and initiatives in which church actors from Germany participate.

"Climate of Justice"—a platform for climate and development

The platform for climate and development was initiated in 2007 when several German churches, mission agencies, and development services started to engage in the Climate Alliance,[18] a broad coalition of churches, trade unions, and environmental and development organizations for climate protection in Germany. Christian stakeholders within this alliance felt the need to agree upon common guidelines and positions for their engagement in climate justice. For this purpose they initiated the platform "Climate of Justice," which enables them to learn from one another and to promote common positions. The signatories to this platform consider poverty reduction, food security, and climate protection as challenges linked to one another and formulate corresponding demands directed toward politics but at the same time toward the churches themselves.[19] In annual conferences, actual challenges are discussed and joint projects are coordinated, for instance, the 2015 Pilgrimage of Justice and Peace to the UN Climate Change Conference in Paris.

Pilgrimage for Climate Justice to the UN Climate Change Conference in Paris, 2015

Taking up the impulse of the 10th general assembly of the WCC, members of the climate platform and other partners of the ecumenical community are planning a Pilgrimage of Justice and Peace to the UN Climate Change Conference in Paris in 2015 and want to invite others to join.[20] The pilgrimage will focus on climate justice because "the victims of climate change represent the new expression of the poor, the widow and strangers, those for whom especially the love of God applies."[21] By bringing reports and voices of the global South into prominence, the initiators seek to raise awareness that political as well as individual actions of people in the global North have significant impacts on societies in the global South. The pilgrimage will offer the opportunity to join for single sections of the journey, being part of the whole at the same time. Spiritual impulses and encounters make it possible to experience the pilgrimage as a community. Furthermore, people on-site can be involved through workshops on climate-justice issues or public actions. Examples of places engaging in best practice in terms of climate protection will be visited in order to encourage new engagement. In addition, so-called points of pain are supposed to point out where changes as well as political guidance are still needed. A list of political demands for climate justice will be developed, which will be signed by many people during the pilgrimage and be handed over to UN officials in Paris.

International Youth Action Day for Climate Justice

How can the church raise awareness among young people that climate protection is a global task but with differentiated responsibilities and different options for change of behavior and lifestyle? The transcontinental networks within mission agencies like the United Evangelical Mission can be a means to involve young people in many different countries in common actions and communication on climate protection.

In 2012 the first International Youth Action Day for Climate Justice was organized by the Evangelical Church of Westphalia, the United Evangelical Mission, and One World Network (NRW). It takes place every year at the beginning of December at the same time as the UN Climate Change Conferences. The idea is to activate young people in many places of the world to carry out, at the same time, activities on climate and environmental protection. These actions are communicated via social media to the other participating groups in order to create a common feeling of being a member of a worldwide movement. Participating youth groups from Indonesia, Rwanda, Congo, Cameroon, Botswana, Colombia, and Germany started activities and projects such as promoting renewable energies, planting trees, climate concerts, an action refusing to use plastic bags, or taking part in a writing contest. Thus, they send a signal that young people all over the world can contribute to the change which is needed and that they are already acting upon. New cooperation with partners like the North German Mission and the World Student Christian Federation should help to include more countries and target groups.[22]

Creation Day / Creation Time

In Germany, Creation Day is usually celebrated on the first Friday in September. It dates back to an initiative of the Ecumenical Patriarch Dimitrios I. He proposed to mark the first day of September, the beginning of the liturgical year in the Orthodox Church and a day dedicated to creation, as European Creation Day.

In 2007 the Third Ecumenical Assembly of Churches took up the proposal. The European churches recommended that the period from 1 September to 4 October should be dedicated to prayer for the protection of creation, and a return and reorientation to a sustainable, climate-protecting lifestyle. In many places Creation Times and Creation Days have become an integral part of the liturgical year. With a central worship service and materials, the Working Committee of Christian Churches (ACK) in Germany offers suggestions for services, prayer, and contemplation.[23]

On Our Way on the Pilgrimage of Justice and Peace

How can Evangelical Churches in Germany strengthen their commitment to climate justice and meet the requirements of a "turning around" to sustainability? Past experience and impulses from global ecumenism make the opportunities listed below appear as pioneering.

- *Preaching ethics of enough.* An efficient use of energy and resources will not suffice to combat climate change and ensure environmental protection. Beside efficiency strategies, sufficiency strategies and alternatives to the current growth model, based on profit increase, material consumption, exploitation, and unregulated global markets are necessary. Churches should introduce to societal discourse the perspective of an ethic that preserves the limits of creation and try to practice sufficiency in their ways of living.
- *Strengthening ecumenical cooperation.* The search for common positions and mutual learning through ecumenical cooperation strengthens the global engagement for climate protection. The call of the WCC assembly in Busan for a seven-year pilgrimage of justice and peace also offers links to the topic area of the integrity of creation. The pilgrimage of

justice and peace should consider climate justice as a focal point. In the ecclesiastical as well as in the nonecclesiastical world, clear messages for ecological justice should be sent out as an ecumenical project. New ways to reach integrity of creation should be tested out, practically and spiritually, and joint objectives should be set.

- *Creating learning places for change.* Through practical climate-protecting measures and by experimenting with a renewed spirituality and attitude in regard to creation, churches can show that a "turning around" is possible. The pilgrimage of justice and peace can be regarded as a frame in which churches set up new targets every year that gradually make them a place of climate protection.
- *Involving young people.* Young people should be engaged to commit themselves to the integrity of creation or their existing commitment should be strengthened. Churches ought to offer projects and participatory opportunities for young people to play an active part in favor of the integrity of creation. Through the global networking of young people (such as that realized at the International Youth Action Day), they can state the common but differentiated responsibility for climate protection.
- *Forming civil-society alliances.* Churches are part of civil society. Cooperation with civil-society actors—such as environmental organizations and human-rights organizations, youth associations, and municipalities at different levels—makes it possible to get a hearing for climate protection in societal and political debates and to achieve synergies concerning the implementation of projects. The cooperation with relevant organizations and networks and the realization of joint projects ought to be enlarged as there are joint projects of church and municipality for climate protection.
- *Taking political action.* Together with other civil society groups, initiatives, and networks, churches should increasingly use their influence on various political levels (UN, EU, nation-state, federal states, and municipalities) for climate protection and climate justice—for instance, at the UNFCC negotiations, in decision making, and to the application of climate-change objectives and measures on national, regional, and municipal levels.

- Besides this, the Sustainable Development Goals (SDGs) offer new opportunities for anchoring sustainability on different levels—for example, regarding the implementation of human rights, the progress of the *Energiewende*, social-policy reforms, as well as the redefinition of prosperity and growth in politics. Churches ought to campaign for strong SDGs and a corresponding implementation in their respective countries.

CHAPTER 11

Deforestation, Poverty, and Climate Change

Evaluating Strategies for Sustainable Living

Lucas Andrianos

One-third of the world forest has been destroyed, a frightening scenario perpetrated by the increase of carbon-dioxide emissions. Despite technological and financial miracles in the 21st century, the gap between the poor and the rich is increasing steadily. "Hungry" people cannot and "poor" communities will not protect their forest areas unless the globalized monetary system will secure their basic needs. Changing strategies for a more sustainable lifestyle are therefore more urgent than ever.

Deforestation, poverty, and climate change are deeply interlinked. Every single issue has to be solved with a holistic approach to the root causes of all problems. There is a global sense that structural greed is the main cause of the current crisis, so we need to develop tools for assessing greed and to measure our progress toward sustainable development.

In this essay, I present methodologies for evaluating a sustainable lifestyle within the limits to greed/growth in order to foster sustainable policies.

Scientific models have been developed to measure progress toward sustainable development (SAFE model) and to assess the structural-greed index on individual and national levels (GLIMS model).

Following the call of the "Green" Ecumenical Patriarchate Bartholomew, the Institute of Theology and Ecology of the Orthodox Academy of Crete (OAC) has repeatedly stressed the need to tackle the ethical roots of the global crisis. Religion and science have to join efforts, and interfaith initiatives such as the Working Group on Climate Change (WGCC) of the World Council of Churches (WCC) should play a greater role in building climate-change awareness. The OAC contributes to this goal by promoting dialogues on environmental ethics, poverty reduction, and climate-change awareness worldwide. It runs academic workshops on greed and the Museum of Cretan Flora for educational purposes.

Situation and Methodology

Since the publication of the Brundtland Report in 1987[1] the concept of sustainability has gained increasing attention among policymakers and individuals. There is clear evidence that development is currently unsustainable. Deforestation, ozone depletion, global warming, depletion of aquifers, species extinction, collapse of fisheries, soil erosion, and air pollution are among the obvious signs of ecological distress.[2] For the last few decades our society is also showing signs that such ecological distress is leading to increased levels of poverty, illiteracy, AIDS, social and political unrest, and violence.[3]

Global deforestation sharply accelerated around 1852.[4] It has been estimated that about half of the earth's mature tropical forests—between 7.5 million and 8 million square kilometers (2.9 million to 3 million square miles) of the original 15 to 16 million square kilometers (5.8 million to 6.2 million square miles)—that until 1947 covered the planet have now been destroyed. Some scientists have predicted that unless significant measures (such as seeking out and protecting old-growth forests that have not been disturbed) are taken on a worldwide basis, by 2030 there will only be 10 percent remaining, with another 10 percent in a degraded condition.[5] This reality is alarming on a local as well as on a global level.

Recently, fuzzy logic[6] has been proposed as a systematic tool for assessing progress toward sustainable development. A model called SAFE[7] (Sustainability

Assessment by Fuzzy Evaluation), which uses basic indicators of environmental integrity, economic efficiency, and social welfare as inputs and employs fuzzy-logic reasoning, was developed to provide sustainability measurements on the local, regional, or national levels.[8] Recently, a similar project entitled GLIMS (Greed Lines and Indexes Measurement System) was also developed by the Greed Line study group of the World Council of Churches (WCC) in order to study the limits and levels of greed in contemporary societies.[9]

This chapter provides an approach to sustainable decision making using sensitivity analysis of the GLIMS model for measuring greed. Sensitivity analysis can reveal the most important factors contributing to a sustainable lifestyle within the limits to greed. Here, I first give an introduction to the concept of sustainable policies and climate-change factors. Next I provide an overview of the SAFE model and its importance in linking deforestation, poverty, and climate change for sustainable decision making. The concept of the GLIMS model and its potential to measure greed for purposes of sustainability and climate mitigation follows. Building on that, I discuss the possibility of evaluating strategies for a sustainable lifestyle with GLIMS computation results on individual levels. This section answers critical questions such as "How greedy is our lifestyle?" and "How does it affect climate change?" To conclude, I report on the activities of the Institute of Theology and Ecology at the Orthodox Academy of Crete[10] in tackling ecological problems and, especially, climate change.

The SAFE Model: Linking Deforestation, Poverty, and Climate Change

Description of sustainable development concept: The SAFE model

Sustainable development, as defined by the Brundtland Report, is the "development that meets the needs of the present without compromising the ability of future generations to meet their own needs."[11] Sustainable development is difficult to define but many researchers recognize that it is a function of two major components, ecological and human.[12]

According to the SAFE methodology, the overall sustainability—Climate Change Risk—of the system whose development we are asked to appraise has two major dimensions: *ecological sustainability* (ECOS/Deforestation) and *human sustainability* (HUMS/Poverty). The se will be referred to as the primary components of the *overall sustainability* (OSUS/Climate Change). The ecological dimension of sustainability comprises four secondary components: water quality (WATER), land integrity (LAND), air quality (AIR), and biodiversity (BIOD). The variables describing the human dimension of sustainability are political aspects (POLIC), economic welfare (WEALTH), health (HEALTH), and education (KNOW).

Figure 11.1 illustrates all the dependencies of sustainability components.

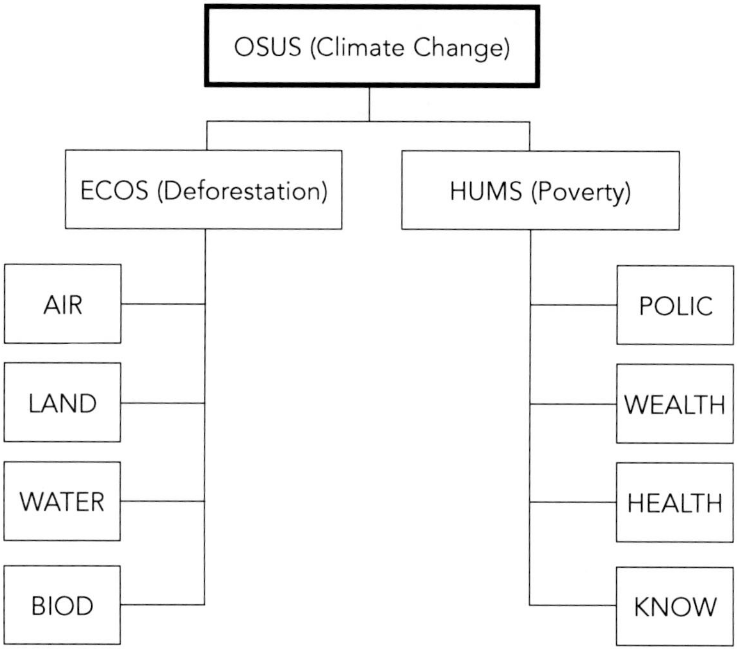

Fig. 11.1. Dependencies of sustainability components

To evaluate the secondary components we adopt the Pressure-State-Response approach, which was originally proposed to assess the environmental component of sustainability.[13] The indicators used in the SAFE model are given in table 11.1.[14] Statistical data for the basic indicators can be obtained from many sources, such as United Nations organizations, the World Bank, the World Resources Institute, international federations, governmental and nongovernmental organizations (NGOs), and so forth.

Basic indicators* used in the SAFE model			
Secondary Compo-nent	PRESSURE	STATUS	RESPONSE
LAND	(1) Commercial energy use (2) Solid and liquid waste generation (3) Nuclear energy production (4) Population density	(5) Net energy imports (6) Domesticated land (7) Forest and woodland area	(8) Population growth rate (9) Primary (clean) energy production (10) Nationally protected area (11) Urban households with garbage collection
WATER	(12) Water pollution (13) Urban per capita water use (14) Freshwater withdrawals	(15) Annual internal renewable water resources	(16) Percent of urban wastewater treated
BIOD	(17-19) Threatened plant, fish, and mammals species (20) Threatened frontiers forest	(21–23) Total number of plant, fish, and mammals species (24) Current forest	(25) Protected area (26) Annual deforestation—reforestation
AIR	(27) CO_2 emissions (28, 29) Total CH_4 and N_2O emissions	(30–34) Atmospheric concentrations of greenhouse and ozone-depleting gases	(35) Fossil fuel use (36) Primary electricity production (37) Public transportation
POLIC	(38) Military spending (39) General government consumption (40) Murders (41) Human rights (42) Environmental laws and enforcement	(43) Regime (democratic-nondemocratic) (44) Institutional Investor Credit Rating (45) ICRG risk rating (46) Central government finance	(47) Official development assistance (48) Government total expenditure for social services

WEALTH	(49) GDP implicit deflator (50) Imports (51) Private consumption	(52) Total external debt (53) GINI index (54) GNP	(55) Resource balance (56) GDP growth (57) Exports (58) Poor households
HEALTH	(59, 60) Cases of infectious diseases: measles and tuberculosis (61) Infant mortality rate (62) Maternal mortality rate	(63) Life expectancy (64–66) Percent of one-year-old infants immunized against measles, polio and DPT (67, 68) Number of people treated per doctor and per nurse	(69) Public health expenditure (70) Daily per capita calorie supply (71) Access to sanitation
KNOW	(72) Number of patent applications filled by nonresidents (73) Lack of libraries/ schools	(74, 75) Expected years of schooling, male and female (76, 77) Gross school enrollment ratio: primary and secondary	(78) Public expenditure on education (79) Number of patent applications filled by residents (80) Personal computer (81) Internet hosts (82) Number of scientists and engineers

*Sources and explanations for indicators in World Bank, World Resources Institute, and the International Helsinki Federation for Human Rights (IHR).

Table 11.1: Indicators used in SAFE model

Sustainable decision making using the SAFE model—sensitivity analysis

Sensitivity analysis plays a fundamental role in decision making because it determines the effects of a change in a decision parameter on the system performance. Additionally, since most decisions regarding sustainable development involve groups of experts, politicians, and individuals, often with uncertain criteria and conflicting interests,[15] sensitivity analysis can be used to investigate the dependencies of sustainability components on particular policies and decisions.

To be able to design policies for sustainable development, one should have a tool for measuring sustainability and a tool for simulating sustainability scenarios. Without these tools, it is useless to formulate any policy for sustainable development, because not only is there no alternative way to assess the results of the policy, but also it is impossible to tell whether the society is on a sustainable path or not.

The SAFE model provides these prerequisite tools for the formulation of sustainable policies by assessing sustainability for different scenarios of development. A scenario is defined by a suite of sustainability indicators, which largely reflect the results of policies and actions taken in a particular period. When these values are changed and the resulting changes on sustainability are observed we can identify the most important indicators promoting or impeding progress toward sustainable development. The next step is to recommend future policies and actions that would increase or decrease the values of the indicators identified as promoting or impeding, respectively among others, the results from the sensitivity analysis of the SAFE model which show that deforestation and poverty are crucial factors influencing overall sustainability and shaping climate change.[16]

Deforestation is a contributor to global warming, and is often cited as one of the major causes of the enhanced greenhouse effect. Tropical deforestation is responsible for approximately 20 percent of world greenhouse-gas emissions. According to the Intergovernmental Panel on Climate Change (IPCC), deforestation, mainly in tropical areas, could account for up to one-third of total anthropogenic carbon dioxide emissions. Reducing emissions from deforestation and forest degradation (REDD) is a mechanism that has been under negotiation by the United Nations Framework Convention on Climate Change (UNFCCC) since 2005, with the twin objectives of mitigating climate change through reducing emissions of greenhouse gases and removing greenhouse gases through enhanced forest management in developing countries.[17]

Overview of the GLIMS Model

Combining religion and science to tackle greed

Greed is the greatest of all plagues against justice, peace, and sustainability. The society of the 21st century relies on unrestricted structural greed and promotes it through unlimited growth, overconsumption, and individualistic competitive behavior. We need an empirical tool to measure, monitor, and control the various aspects of human greed on global, national, corporate, and individual levels. Jesus Christ reminds us to "take care to guard against all greed, for though one may be rich, one's life does not consist of possessions" (Luke 12:15, New American Bible Revised Edition). Religion and science should give answers to critical questions such as "What is greed?" and "How can we measure and control it?"

The conceptual description of structural greed is driven from the findings of the WCC greed-line group study on poverty, wealth, and ecology.[18] Fuzzy-logic evaluation is used to assess greed.[19]

The concept of greed and its measurement: The GLIMS model

Greed could be defined as the *desire to have more than one's legitimate share of material goods and power.* In contrast to poverty, which deals with needs that can be objectively defined and even quantified, greed is about desires which are "difficult to contain," involve an "emotional energy that seeks to transgress or disregard limitations," and which are consequently difficult to circumscribe and measure.[20]

If greed is "having too much" money, resources, and power (in contrast to describing poverty as" having too little"), when does one "have too much"?[21] It was proposed that the point or level at which individuals or societies "have too much" is approached or describes a situation or status, first, when other individuals and societies have too few resources to live by, and, second, when the accumulation of wealth and power undermines the common good or threatens (desire/trends) the global commons.

While the "poverty line" is drawn at the point of personal consumption, allowing for the satisfaction of basic needs, the greed line could be drawn at "the highest point of personal consumption which can be obtained without negatively affecting the welfare of society and that of future generations."[22] "Greed lines" are the levels of resource consumption, money accumulation, or power seizure over which societal or individual behaviors may harm human well-being and creation integrity. These negative effects of behaviors beyond greed lines could be expressed in term of relative poverty or socioeconomic injustice or sentimental offenses and environmental destructions.

The GLIMS model uses fuzzy-logic reasoning and inputs from statistical indicators of natural-resource consumptions, financial realities, economic performances, and social, ethical, and political facts. The outputs are concrete measures of three primary indexes of ecological, economic, and sociopolitical greed (ENV-GI, MON-GI, POW-GI) and one overall multidimensional structural greed index (MSGI). The results are greed-index scores that are expressed on a scale of zero to one hundred. The global greed-index score equal to 100 corresponds to the maximum level of greed for the subject of analysis.

In contrast to the poverty line, the GLIMS model allows the formulation of discrete limits to all aspects of greed and it gives an overall multidimensional structural-greed index (MSGI).

The knowledge ruling the computation of greed indexes is represented by fuzzy rules whose general form is: "*If* (PREMISE) *then* (CONCLUSION)."

For example, such a rule could be as follows:

- "*If* (POVERTY is *high*) *and* (ECOLOGICAL FOOTPRINT/Deforestation is *high*) *then* (OVERALL SUSTAINABILITY/Climate change risk is *high*)"; or
- "*If* (POVERTY is *medium*) *and* (ECOLOGICAL FOOTPRINT/Deforestation is *medium*) *then* (OVERALL SUSTAINABILITY/Climate change risk is *medium*)"; or
- "*If* (POVERTY is *low*) *and* (ECOLOGICAL FOOTPRINT/Deforestation is *low*) *then* (OVERALL SUSTAINABILITY/Climate change risk is "*low*"); etc.

The rules are expressions of the role of interdependencies among various dimensions of greed. They are combinations of IF-THEN rules operating on rule bases derived from expert knowledge of the system integrity. By their nature, such functions are highly nonlinear. The term *integrity* is defined as the degree to which each greed variable fulfills criteria of greed lines. Criteria of greed lines are recommended critical targets that each greed indicator should pass to become in a greedy status. These rules are the results of multidisciplinary analysis about greed and its polymorphous effects.

Table 11.2 summarizes the list of greed indicators and their limits to greed used in the GLIMS model.

Greed indicators	Personal /institutional data values			
	Legitimate target	Greed line interval		Greed line unit
1. PERSONAL ECOLOGICAL FOOTPRINT	Lower better Red light when ratio is > 1 Earth.	0.9	1	1= one planet Earth
2. FINANCIAL ASSETS	Moderation Red light when ratio is > 100%.	80	100	Percentage of revenue
3. FINANCIAL INTEREST RATE	Moderation (5% is the world average interest rate) Red light when ratio is > 5%	0	5	Percentage
4. ANNUAL IN-COME	Moderation Red light when ratio is > 40000 USD $ /capita/year	30000	40000	Money in USD $ based on purchasing power parity (PPP)/ capita)
5. HOUSEHOLDS DEBT	Moderation Red light when ratio is > 100%	80	100	Percentage of revenue (installment vs. income ratio).
6. LIVING STAN-DARD COSTS	Moderation Red light when ratio is >0,8	0,6	0,8	In a scale of zero to 1
7. SOCIAL CLASS STATUS	Moderation Red light when ratio is >0,8	0,6	0,8	In a scale of zero to 1
8. EDUCATIONAL EXPENSES	Moderation Red light when ratio is >0,8	0,6	0,8	In a scale of zero to 1
9.HEALTH CARE EXPENDITURE	Moderation Red light when ratio is >0,8	0,6	0,8	In a scale of zero to 1
10. PERSONAL MORALITY	Higher better Red light when ratio is <0	0	5	In a scale of -10 to +10
11. RESPECT FOR PERSONAL FREE-DOM	Higher better Red light when ratio is <5	5	6	In a scale of 1 to 7

Greed indicators	Indicator definition and link to greed	Data sources
1. PERSONAL ECOLOGICAL FOOTPRINT	Amount of biologically productive land and sea area necessary to supply natural resources and to assimilate associated waste for a given lifestyle.	Personal ecological footprint can be calculated at http://www.footprintnetwork.org
2. FINANCIAL ASSETS	Amount of money that an individual has in the financial market.	Questionnaire
3. FINANCIAL INTEREST RATE	Depending on the desire for making more money in a short time, a higher interest rate is considered to be result of greedier behavior.	Questionnaire
4. ANNUAL INCOME	Higher personal revenue compared to the average population is a sign of greed.	Questionnaire
5. HOUSEHOLDS DEBT	Because of imbalanced behavior (production vs. consumption), the limit over which a person or an institution is considered as greedy can be set to 100%.	Questionnaire
6. LIVING STANDARD COSTS	Access to adequate living standard is a fundamental need for human right but luxury living is offensive and becomes a form of greed when in excess compared to the average standard.	Questionnaire

Greed indicators	Indicator definition and link to greed	Data sources
7. SOCIAL CLASS STATUS	High level of social class distinction is a form of greed with respect to dignity and inequality when compared to the average standard.	Questionnaire
8. EDUCATIONAL EXPENSES	Education a fundamental need for human right but it becomes a form of greed when educational expenses are in excess compared to the average standard.	Questionnaire
9.HEALTH CARE EXPENDITURE	Adequate health is a fundamental need for the fulfillment of human right but it becomes a form of greed when health expenditure is in excess compared to the average standard.	Questionnaire
10. PERSONAL MORALITY	Morality index intends to provide a time-base measure of the moral behavior of individual. The red light is 0, under which the bad behavior is threatening human life and peace.	http://www.moralityindex.com/ Questionnaire
11. RESPECT FOR PERSONAL FREEDOM	One should not tolerate any violation of freedom as it is an obvious sign of greed. Higher values reflect high level of liberty and low values imply violation of human right in a greedy society. Values range between min=1 and max=7.	Questionnaire

Table 11.2: Explanations for the greed indicators on individual level

The methodology of the measurement of greed indexes, as well as the concept for the selection of greed indicators are depicted in the following figure (fig. 11.2).

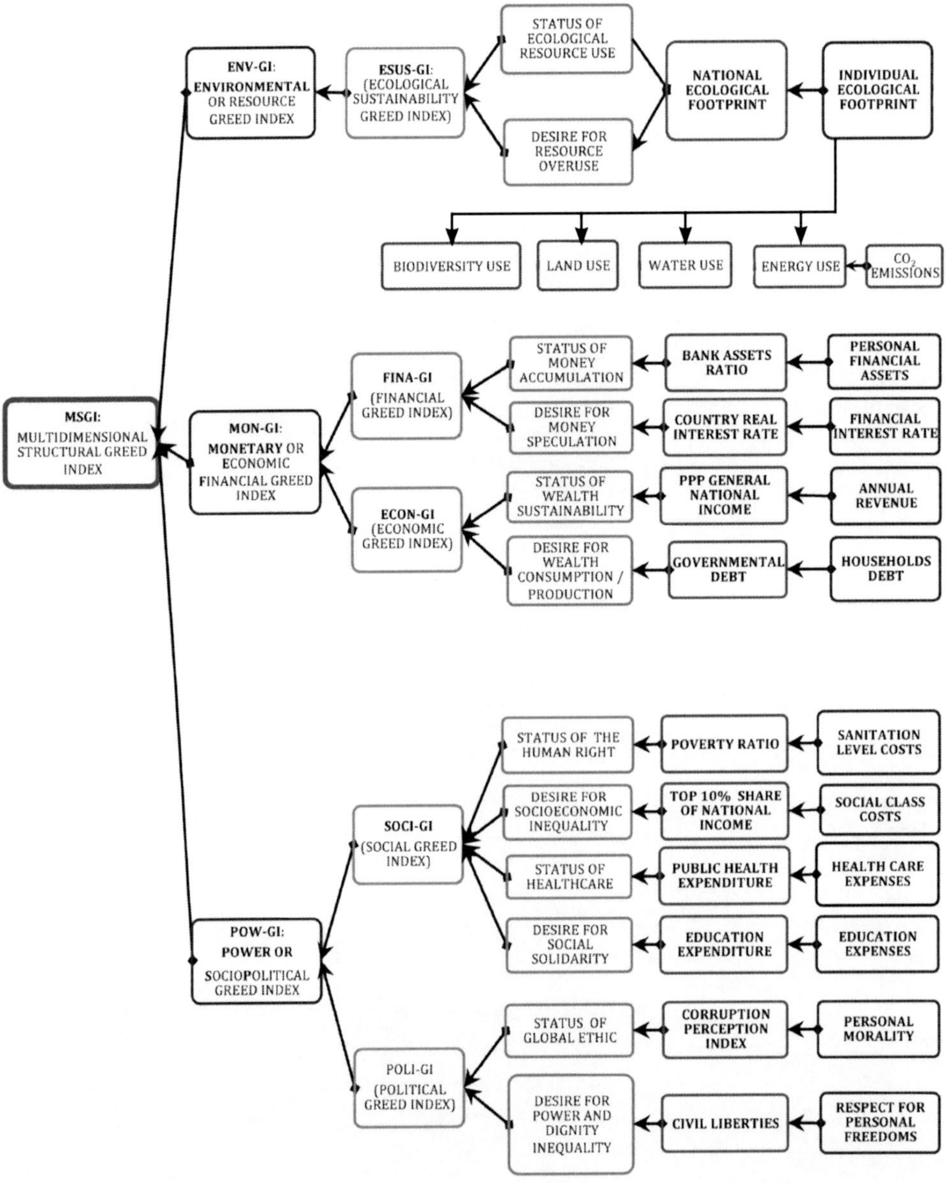

Fig. 11.2. Greed indexes and indicators

Evaluating Strategies for Sustainable Individual Lifestyles with the GLIMS Model

Evaluating strategies for a sustainable lifestyle on individual levels is a pilot project of the WCC and the Institute of Theology and Ecology of the OAC in order to promote sustainability and to avert climate change. To evaluate individual greed indexes, we applied the GLIMS model to selected economies and a few individuals.

One individual is considered greedy with respect to a specific "dimension" of structural greed when its greed index is greater than 50. The greed line for MSGI measurements is then 50 to 60 and the red light is 40 (the "enough" level). Data for the whole set of greed indicators for countries are retrieved from statistical researches while data for individuals are obtained via surveys that are based on questionnaires such as follows:

Box 11.1. Personal Greed Index Questionnaires

1. How big is my ecological footprint (to calculate, go to http://www.footprintnetwork.org/en/index.php/GFN/page/calculators/)?
2. Do I have stock shares or savings capital in the bank? If yes, how much?
3. If I have financial investments, how much is my financial interest rate?
4. How much is my annual income? And how high/low it is compared to the average?
5. How much credit card debt do I have? How much is my monthly installment vs. my monthly income?
6. How much do I spend on luxury goods in a year? How many cars do I have? How big is my house?
7. Would I consider myself to be part of the upper, middle or lower income class? Do I aspire to a higher social class? If so, which class?
8. How much do I spend for my health care (except for diseases and disabilities)?
9. What level of education do I have (in years of schooling)? How much do I spend on my education per year with respect to the average neighbors?
10. How much tolerance do I have for corruption and for human rights violations?
11. How much do I value money and power over friendships, sharing, harmony, and wisdom?

Here we report the results of individual assessments which show that the environmental, economic, and social greed levels of each person would need corrections because they have all over passed the red-light level of "enough" (MSGI equal to 40). The results are compiled in the following table (table 11.3):

INDIVIDUAL GREED INDEXES	Person 1 N (Male)	Person 2 L (Male)	Person 3 A (Female)	Person 4 G (Female)	Person 5 S (Young)
ESUS-GI	40.26	41.28	55.29	50.41	44.28
ENV-GI	40.37 (E)	41.53 (E)	54.90 (G)	50.36 (G)	44.77 (E)
FINA-GI	20.37	22.80	56.91	20.37	20.37
ECON-GI	46.20	50.41	41.69	36.93	26.05
MON-GI	38.31 (F)	40.53 (E)	49.79 (E)	35.48 (F)	32.19 (F)
SOCI-GI	43.17	53.74	55.82	44.31	43..40
POLI-GI	37.29	32.08	59.21	37.28	28.63
POW-GI	44.33 (E)	45.43 (E)	51.88 (G)	44.69 (E)	41.04 (E)
Overall MSGI score	45.50 (E)	46.23 (E)	50.37 (G)	46.73 (E)	44.94 (E)
Linguistic values for greed indexes: L = Low; F = Fair; E = Enough; G = Greedy; VG = Very Greedy; EG = Excessively greedy					

Table 11.3: Values of greed indexes measurements for selected individuals (2012 survey)

Other Activities of the Institute of Theology and Ecology at the OAC

Sustainable Alternatives for Poverty Reduction and Ecological Justice (SAPREJ)

In April 2014, the WCC and the Orthodox Academy of Crete organized the second international conference on "Sustainable Alternatives for Poverty Reduction and Ecological Justice" (SAPREJ) in Antananarivo Madagascar. The first SAPREJ conference was held in Greece in 2012 at the Orthodox Academy of Crete in Kolympari, Crete.[23] Greece is most affected by the global financial crisis and Madagascar is among the poorest countries in the world. Poverty has increased and deepened substantially over the last two-and-a-half decades, with real per capita income having decreased by 40 percent between 1971 and 1991. The poverty assessment for Madagascar estimates that 70 percent of the population can be defined as being poor and 59 percent as being extremely poor. Poverty is also deeper in rural areas than in urban areas, with an unemployment rate of about 80 percent. Madagascar has lost 80 percent of its eastern rainforests since then.[24]

Responding to the call for a pilgrimage of justice and peace at the 10th general assembly of the WCC in Busan, South Korea, the OAC conference was meant to bring a message of hope and solidarity to the needy and marginalized communities in Africa. About 25 proposals from a dozen countries (Canada, Germany, Greece, Fiji, Qatar, USA, Nigeria, Uganda, Zambia, Algeria, The Netherlands, and Madagascar) were presented.

Madagascar and the global South (developing countries) in general expressed their call for justice and mutual collaboration with the global North (developed countries). They asked for the support of the WCC, NGOs, and all churches in solving the problem of climate change through eco-justice and poverty-reduction programs.

Ecological Theology and Environmental Ethics (ECOTHEE) and Museum

In 2008, the Institute of Theology and Ecology at the OAC started to organize international conferences, seminars, and workshops focused on ecological theology and environmental ethics (ECOTHEE)[25] and biodiversity conservation (Museum of Cretan Flora).[26] Each ECOTHEE conference takes place every two years at the Orthodox Academy of Crete where priests, scientists, politicians, and the general public come to reflect together on finding solutions to ecological problems. An ECOTHEE book series, resulting from the outcome of each conference, has been published and distributed worldwide.

The museum of the OAC exhibits about 9,000 specimens of Cretan wild plants and offers various educational programs demonstrating the interrelation between plant conservation and climate-change mitigation. Thousands of tourists, scientists, students, church members, and the general public visit the museum every year. Two other important international conferences on plant biodiversity were also organized by the OAC in 2010, for the Conservation And Sustainable Use of Wild-Plant Diversity (CSUWPD), to celebrate the year of biodiversity, and in 2014, for the conservation of wild plants in Europe to mark the 7th Planta Europa conference in Crete. During these conferences, the issue of climate change is always at the center of discussions.

Since 2011, the Institute of Theology and Ecology at the OAC has been organizing workshops and seminars on the limits to greed (greed line) and sustainable lifestyle management. Lectures on structural greed have been given worldwide to spread the call for a change in lifestyle with respect to the limits to greed/growth.

Conclusions

Church leaders, policymakers, and all believers need a practical tool to clarify the effects of greed and establish policies or strategies for a sustainable lifestyle. With the help of science and religion, we have developed tools for monitoring greed and sustainable development. The SAFE (Sustainability Assessment by Fuzzy Evaluation) model allows decision makers to assess their progress toward sustainable policies while the recent model, called GLIMS (Greed Lines and Indexes Measurements System), is meant to provide a comprehensive description and measurement of structural greed. The GLIMS model offers an explicit way to assess greed and help shape sustainable policies to tackle climate change at its roots (poverty, ecological footprint, deforestation, ethical recessions, education, and inefficiency in production and distribution systems). The model allows also the monitoring and simulation of greedy behaviors at the national, corporate, and individual levels. Therefore, it is helpful for antigreed policy making and ethical reflections.

It is crucial to understand that poverty, deforestation, and climate change are interlinked and governed by structural greed. To overcome greed, recommendations differ from economy to economy and corrections vary from individual to individual. More-developed countries need to focus on the effect of their environmental greed, whereas less-developed countries should strive to correct their sociopolitical systems. The monitoring of sustainability and the greed index are indispensable for decision making and policy design at individual, communal, or national levels if we want to attack the problem of climate justice, peace, and sustainability more efficiently.

CHAPTER 12

Eco-Theology, Climate Justice, and Theological Education

Raphael Sartorius and Dietrich Werner

Curriculum Development on Eco-Theology and Climate Change

During the last World Council of Churches (WCC) assembly in Busan, South Korea, in November 2013, Ecumenical Conversation No. 12, which took place under the theme "The Earth Community Groans: A Call to Ecological Justice and Peace in the Face of Climate Change," four ecumenical confirmations emerged as strategies for the future work of the WCC:

1. We affirm that the crisis that God's creation is facing is fundamentally ethical and spiritual and is a threat to all of life.
2. We affirm that climate change is one of the most threatening manifestations of the ecological crisis affecting all life on the planet, particularly the

most vulnerable everywhere. Too little has been done by the international community.

3. We affirm that scientific data indicates that the ecological crisis, caused by the actions of human beings, is already leading to an irreversible threat to humanity and the entire planet, and thus urgent action is needed.
4. We affirm that some churches around the world are addressing this issue locally, nationally and globally, working also on an ecumenical, interfaith and inter-generational level, but we acknowledge that a lot more needs to be done.

One practical recommendation which followed was: "In order to experience radical change, we challenge the churches and other organizations to raise awareness, do and share good practices and engage in advocacy work."

"Raising awareness" points to the dimension of education, both for laypeople and for the ordained, in which issues of eco-theology and climate justice can deepen their theological presuppositions and implications. During the recent WCC seminar on "Green Churches—Ecology, Theology and Justice in Practice," in June 2014,[1] among other concerns was the issue of eco-theology and climate justice in theological education and curriculum plans in churches worldwide. Participants stated that this is far from being satisfactory and not consistent in certain regions. While there are some regions like India, where the Senate of Serampore College has developed a new BD curriculum in which, since 2010, people can find a BD course on green theology; and some universities, like the University of the Western Cape in South Africa, which has offered courses and expertise on ecological ethics and eco-theology for several years already,[2] other regions and schools have no research institute or regular curriculum plan on eco-theological developments.

A recent joint global research project on trends and challenges in theological education in world Christianity (global survey on theological education), which was based on digital questionnaires and answered by more than 1,700 experts in theological education worldwide, has shown some alarming signs in terms of the stability, accessibility, curriculum deficits, and trends of decline in theological education. The complex data has been summarized in a short research report, which was introduced during one of the ecumenical conversations at the Busan assembly.[3]

The following results are significant in understanding how far systems of theological education and curriculum development have succeeded or are aware of the need for integrating new eco-theological course outlines in the context of the global ecological crisis:

- There are not enough theological schools in those parts of the world where Christianity is growing rapidly (Africa, Latin America, and parts of Asia).
- Theological education is financially unstable in many parts of the world. Only 2 percent of the respondents from Asia saw theological education in Asia as being stable, while almost 70 percent saw theological education as financially unstable or even in a financial crisis.
- Growth is seen in Evangelical and Pentecostal/Charismatic theological education; there is a decline in mainline Protestant and Roman Catholic traditions.
- Only a minority of those who responded (35 percent) see an important need for additional courses on environmental studies,[4] while a majority recommend courses on cross-cultural communication and practical skills as most urgent to be added.[5]

There is no concise survey on how many of the around 3,000 theological schools worldwide actually have implemented courses on eco-theology already. The seminar on "Green Churches" in Bossey in June 2014 suggested that a "Global Green Churches Handbook" should be developed, which would bring together crucial theological perspectives, checklists, and best-practice models for churches, and which could be also used as a resource book for theological education. Apparently, a vast need still exists for sharing innovative and encouraging models and to get them rooted not just as an option but as an obligatory course in theological education.

How to Make Your College Green: The Green Seminary Initiative

A significant project for the systematic turn toward sustainability within institutions of theological education is introduced here and represented by the Green Seminary Initiative (GSI). This initiative is an outstanding example from the U.S. region.

The Green Seminary Initiative (GSI) was launched in 2007. It was carried out by GreenFaith, an interfaith coalition for environmental issues. Two major visions inspire the GSI: (1) the desire, as part of the religious community, to help restore areas that have experienced the humanmade environmental destruction which affects all creation as well as (2) to provide theological seminaries and universities with the necessary knowledge to teach their students—future key leaders of congregations, Christian communities, and organizations—on how to preserve and restore all creation in an holistic approach. Therefore, GSI aims to encourage theological schools to incorporate care for creation as part of their curricular models and into their mission statements. The GSI is directed by a steering group consisting of numerous North American scholars, pastors, and organizational leaders.[6] The GSI serves as a hub to achieve the mentioned goals, namely, to provide knowledge—such as academic syllabi on eco-theology—and strategies to share models on eco-justice and sustainability, both within academic programs as well as school facilities. All in all, the GSI homepage provides rich resources of great scale.

This essay will focus on two aspects: first, a summary of the GSI-survey project; and, second, a report about findings about what obstacles remain for a fully integrated eco-theology program in church-related seminaries, followed by an introduction to the Green Faith Certification Program.

In 2013, the GSI Survey Project[7] was sent to 27 theological schools or schools containing a faculty of religion within the United States, and an informational website was established. These schools were initially preselected with the intention of having an adequate range of different theological orientations. The individuals most concerned with care-for-creation issues were identified and some of them were later interviewed. Nineteen schools responded to the survey. The primary information gathered by the survey was built from the following five questions:

1. To what extent is an institutions' curriculum already emphasizing the significance of creation and today's ecological challenges?
2. In what way are facilities and housing fulfilling sustainable and environmental criteria on an operative and constructional basis?
3. Are community life, public advocacy, food services, outreach, and mission subject to the concern for creation?
4. Are key stakeholders (students, board, alumni, funders, staff, etc.) integrating care for creation into their corresponding operational business?

5. Would the GSI be able to support these efforts through its website, conferences, or any other measures?

The report provides a wide range of possibilities to encourage care for creation in theological education and can be summarized in the following points.

First, the GSI observed that care for creation is still an add-on in most of the schools and not included in their mission statements. None of the observed schools had a developed and fully integrated care-for-creation program. It was noted that the success of such a program is always dependent upon a key figure in the school who is constantly promoting and pushing the idea. Most existing care-for-creation programs were started by faculty, possibly due to the fact that student initiatives are more difficult to sustain.

What are major obstacles to building better-developed care-for-creation programs, whether in the academic curriculum or in improving school facilities? The report mentions several factors:

- care for creation is still seen as something extra and not essential;
- the idea is not institutionalized thus far;
- some schools have declared environmentalism as being too liberal or even pantheistic;
- environmental concerns related to the renovation of old buildings were rejected because they were seen as too costly;
- several faculty members admitted having a lack of knowledge on how to incorporate care for creation in their respective disciplines.

As regards opportunities, one or more schools identified the possibility[8] of connecting with other schools and universities to establish multidisciplinary relations (for example, medical or forestry schools). New sustainable buildings could fuel inspiration toward new program opportunities. Tying the care for creation with social justice was mentioned, as well as grants and a relationship with GreenFaith.

Since then, GreenFaith and the GSI have conducted further research on seminary environmental efforts. Interviews were conducted with faculty, administration, and students at over 30 seminaries. In addition, a detailed review was conducted of the 25 seminary environmental profiles that were submitted to GSI. In-depth personal interviews were conducted with representatives of a cross-section of schools based on denominational affiliation, size, location, and theological orientation (conservative to progressive).

In 2015, the GreenFaith initiative will launch the "GreenFaith Certification Program." This certification process lasts three years and was developed particularly for seminaries, divinity schools, theological schools, and houses of worship. Greenfaith provides structured support. Seminaries are to educate their students on sources from sacred texts, theology, and ethics that speak about humanity's right relationship with creation. They should equip their students to teach, lead worship, and help others develop spiritual practices that strengthen their community's relationship with the natural world. This includes one-on-one coaching, support in managing properties, and the training of staff in an environmentally sustainable manner to integrate care for the creation into the ethos of their institution through worship, food services, community and social events, alumni relations, and other means, in order to successfully complete the certification process and become a green institution.[9]

A Worldwide Mapping of Resources: The Global Survey on Ecotheology, Climate Justice, and Food Security[10]

One tool to enhance global networking and mapping on resources for eco-theological education has been recently developed by a joint initiative and project between the WCC, Globethics.net (Geneva), and Bread for the World, Berlin. The WCC assembly in Busan identified eco-theology and climate justice as two of the key priorities for the next working period (2014–2021). Issues like eco-theology and climate justice will be among the key components in the global pilgrimage of justice and peace. Since climate change and ecological destruction are getting increasingly menacing for humanity, climate issues provide a global challenge which should be reflected also in theological education and formation worldwide. Specialized ministries and their partners in the global South have shown the links between climate change and food security. Churches, ecumenical organizations, specialized ministries, and institutions for theological education and research worldwide will be asked to contribute to the pilgrimage of justice and peace, as no major working project in the WCC in the post-Busan working period will be possible without vital strategic cooperation, connectivity, and common ownership with ecumenical partners in the global ecumenical movement.

According to preliminary data, although much research and project-related work has been done already on these topics in some regions, only in a few institutions of theological education and Christian leadership development do they form part of the regular curriculum of instruction and training or inform theological formation processes in an integrated perspective,[11] especially in exchange with the global South. A great potential therefore lies in discovering existing resources for the transfer of knowledge and wisdom from all contexts/regions. The ecumenical sharing of resources about these issues could be a benefit for all.

The ETCF project has four major aims:

1. Highlighting and visualizing the contributions, resources, and perspectives of eco-theology, climate justice, and food security (ETCF), particularly from the churches of the South, as often resources and contributions are only known and shared in the North.
2. Mainstreaming ETCF as integral components in Christian leadership formation, both in formal theological education as well as in nonformal programs of Christian and development training in the global South.
3. Supporting and fostering the relevance of ETCF in the awareness of churches in both the global North and South in a period when worldwide political negotiations on binding restrictions on CO_2 emissions are reaching a certain climax (2015).
4. Contributing to international networking, exchange, and South-South, North-South, and West-East networking between regional hubs of research and education on issues of ETCF.

The questionnaire[12] seeks to receive answers from theological colleges, church-related seminaries, universities with their faculties of theology or religious studies, and other church-related training facilities and programs. More answers are expected from selected Christian or interfaith research centres, including lay academies working on issues related to ETCF and eco-ethics.

The ETCF project expects to receive results of a well-designed questionnaire, probably followed by in-depth interviews with selected individuals from institutions of theological education, training, and development, as well as regional associations of theological schools. Another intention is to make resources and examples of good practice on ECTF available.

A list of resources identifying monographs, journal articles/essays, key church statements), curriculum models, and training tools (specific courses) which have been developed and practiced in the global South related to ETCF, as well as examples of good practice in relation to ECTF (green churches, eco-congregations, etc.), will be published after the project is finished on the related website of Globethics.net.[13]

Major hubs and research institutes in the global South working and providing education or innovative projects in the area of ETCF will be named and published on a common web platform within the Global Digital Library on Theology and Ecumenism (GlobeTheoLib) on ETCF resources. After the process is completed, a special thematic collection with digitized resources for theological training on ECTF will be launched in 2015 within GlobeTheoLib. It will include a limited list of good-practice examples with curriculum models on ETCF and a limited compilation of core publications and teaching resources, which can be recommended as significant and crucial to highlight the voices of Christian churches from the global South on ETCF.[14] As a benefit, individual persons and institutions of theological education participating in the ETCF project will gain access free of charge to the new ETCF resources website and will be linked to the emerging network of research and educational hubs engaged in ETCF projects.[15]

The publication of the final ETCF survey report will hopefully provide a rich and more accurate picture on how theological schools, particularly in the global South, have started to integrate eco-theological concerns into their own curriculum plans and have produced expertise and resources which should be known more widely in ecumenical partnership between churches and their institutions of theological education. There are more and more theological seminaries taking on issues of eco-theology in order to develop their own profile of doing theology contextually.[16] The growing network of green seminaries and the related resources list of the ETCF project will help to facilitate this process.

CHAPTER 13

Salim, Ever-Recycling Energy

A Korean Postcolonial Eco-Feminist Theology of Life

Jea Sophia Oh

Both colonialism and neocolonialism are about human colonization over other humans. Neocolonialism is often recognized as economic imperialism in the age of a global economy, in which a nation practices capitalistic power, military control, and political control toward other nations. The same colonial mindsets have created similar damaging illusions about human mastery over nature. Now is the time to think about human colonization over nature. This study offers a postcolonial eco-theology on nuclear energy by observing the Fukushima nuclear disaster and its damages and effects on the vegetation and biosphere of Korea since 2011. First, I show that the so-called nuclear renaissance, as an alternative energy to the climate and energy crisis, is a delusion based on human arrogance that humans can exercise control over nature. Thus, it is *jugim* (destructive power) and not *salim* (enlivening power). Second, in order to prevent exhaustion of energy resources, I show that effective recycling is a better solution, along with developing greener forms of energy.

As a constructive practice of recycling, I introduce Korean women's *salim* movements (*ah-na-ba-da* movement), which include saving (*ah*), sharing (*na*), exchanging (*ba*), and reusing (*da*). Although humans think of energy only as solid (ontological) forms of resources that exist for their own use, the term *energy* is also used outside of physics. Energy, not resources, is the source of life. Energy is embodied in nature as the power of life which will never be exhausted, but will be renewable and recyclable. Korean women's recycling movements arise from their *salim* spirituality. Indeed, it is the power of life. *Salim* creates and recreates life abundantly and sustainably in terms of quality, and finally heals and overcomes the brokenness of natural harmony.

Jugim (destroying): Energy and Climate Change

> To meet our growing energy needs and prevent the worst consequences of climate change, we need to increase our supply of nuclear power. . . . But energy leaders and experts recognize that as long as producing carbon pollution carries no cost, traditional plants that use fossil fuels will be more cost-effective than plants that use nuclear fuel. That is why we need comprehensive energy and climate legislation to create a system of incentives to make clean energy profitable.—Barack Obama[1]

Salim is a Korean term, which literally means making things alive, restoring, and enlivening.[2] The antonym of the word *salim* is *jugim* (intentional killing). *Jugim* refers to all destructive activities, such as killing, marginalization, oppression, exploitation, coercion, colonization, contamination of the environment, destruction of the ecosystem, and so forth. *Jugim* is not natural but exists contrary to nature and the unnatural activities of life which engage in violence.[3] Is nuclear energy really a good clean energy of *salim* or a mad, dirty energy of *jugim*? What do the lessons from Hiroshima to Fukushima teach us?

More than two years since the Fukushima Daiichi nuclear disaster that began on 11 March 2011, its impact remains massive and widespread. It will be many decades before the full scope of the impacts of this horrible ongoing disaster is fully understood, but significant health, economic, and environmental consequences are already evident. Not only has this disaster caused significant damage to the environment, the health of populations, and national economies, but the heavy financial cost of a meltdown is inevitably borne by the public, not by the companies that designed, built, and operated the plants. The Fukushima nuclear disaster showed us that nuclear reactors are dangerous and ineffective. But most of all, it is *jugim* (destroying) when radiation effluent comes to reality. It is one of the most horrible disasters this planet has seen.

Nevertheless, the U.S.A. still supports nuclear energy even after the Fukushima disaster. Basically, the Climate Action Plan that President Obama announced in 2014 aims, first, to reduce carbon emissions; second, to prepare the U.S.A. for the impacts of climate change; and third, to lead international efforts to fight climate change and prepare for its impacts. It is seemingly eco-friendly. However, it is not only eco-unfriendly but also imperialistic. The plan states: "In the past three years we have reached agreements with more than 20 countries around the world, including Mexico, South Africa, and Indonesia, to support low emission development strategies that help countries to identify the best ways to reduce greenhouse gas emissions while growing their economies."[4] The United States plans to promote nuclear energy worldwide. Neocolonialism is the name for the political and economic reconstitution of previous colonial subjugation. Robert J. C. Young, a British postcolonial theorist, writes, "Think of the Pentagon and the CIA in Washington, with their global strategy of controlling events in independent states all over the world in order to defeat communism or Islamic resistance and further U.S. interests."[5] Neocolonialism and imperialism are reconstituted in the post–World War II milieu. The American Empire has predominated largely over against a marginalized left. A leading eco-theologian, Catherine Keller, says that the American Empire, through a virtual Wild West of diplomatic vulgarity, cultural homogenization, and maximum firepower, pursues a more global dominance than any previous empire.[6]

The U.S.A. has a large number of bilateral nuclear cooperation agreements with various countries, including South Korea, as specified under Section 123 of the U.S. Atomic Energy Act in 2012. The National Association of Manufacturers and the U.S. Chamber of Commerce and the Nuclear Energy Institute urged the Obama administration to adopt a more determined and pragmatic approach to increasing international trade in nuclear goods and technologies.[7] South Korea is a pro-nuclear government, pressing ahead with a large nuclear program, as is India. According to *Nuclear Power Daily*, in January 2013 the South Korean government granted final approval to build two new nuclear reactors at the Shin Kori Nuclear Power Plant in Ulsan, about 255 miles southeast of Seoul. South Korea's long-term energy plan finalized that the country needs to build at least 12 additional nuclear reactors by 2024 to maintain the proportion of electricity supplied by nuclear reactors at around 30 percent. Construction on five new reactors, including Shin Kori Reactor units 3 and 4, is already under way.[8]

Environmental groups have warned that if the Gori facility fails, people living in this zone could be exposed to extreme radiation, as seen in the two catastrophic disasters. In February 2012 Kori 1 lost all power and the generator did not start. But this serious incident was not reported for a month. Although the government has subsidized nuclear plants, the antinuclear movements and activisms have arisen from environmental groups, religious groups, unions, co-ops, professional associations, and, of course, people living in a dangerous zone. For example, in December 2014 a group of South Korean thyroid-cancer patients living near nuclear plants filed the country's first class-action suit against the operator.[9] Choi Yul, the president of Korea Green Foundation, states, "The Fukushima nuclear disaster has proven that nuclear power plants are not safe."[10] Suh Kyun-Ryul, a professor of atomic engineering at Seoul National University, also says, "The public has totally lost trust in nuclear power."[11] Learning from Hiroshima to Fukushima, it is evident that nuclear energy cannot be a clean alternative power for our planet even though it has relatively low carbon-dioxide emissions. As Alexander Glaser at Princeton Environmental Institute observes, "It will take time to grasp the full impact of the unimaginable human tragedy unfolding after the earthquake and tsunami in Japan, but it is already clear that the proposition of a global nuclear renaissance ended on that day."[12]

Among the antinuclear activists, Korean women are at the frontline. In January 2012, 22 women's organizations in South Korea made a plea for a nuclear-free future. They picketed with signs: "We want a nuclear-free peaceful world!" "Nuclear power reactors are a matter of life or death!" Korean women activists said that they feel an enormous sense of crisis after the Fukushima nuclear disaster, which showed the *jugim* power of radiation in the loss of human lives, pollution of eco-systems, and especially, contamination of foods. An antinuclear activist and a mother, Gyung-Lan Jung, says, "Nuclear reactors threaten our lives, the lives of our children, and all living creatures."[13] It must be a highly sensitive matter to Korean mothers since mothers cook meals for their children. Even my mother in Korea told me that she rinses rice about ten times since she has heard that radioactivity was detected in the rice crops. In order to feed her family, a mother cooks *bab* (rice, the daily meal for Koreans) for her children. Cooking *bab* is also called *salim*.[14] If she feeds her children *bab* with the radioactive rice crops, *salim* transforms into *jugim*.

As I discussed above, there is discrepancy regarding nuclear energy on both sides of the argument. Both the U.S. and the South Korean governments insist that nuclear energy is a clean alternative energy, whereas the Korean ecological activists strongly reject that as *jugim* power. A world-renowned eco-activist, Vandana Shiva, considers "eco-apartheid" as war: "Not only is corporate power converging with state power for the great resource grab, corporate-state power is emerging as militarized power to undemocratically impose an unsustainable and unjust agenda on the earth and its people. That is how the war against the earth becomes a war against people, against democracy and against freedom."[15] Indeed, nuclear energy is a fake green power and human imperialism against the ecosystem. As we have learned from the Fukushima disaster, the state exercises its sovereign power to dominate the nonhuman nature, which has been greatly devastated. The land is made uninhabitable by nuclear disasters. This is a complete catastrophe! Our planet is a living organism that breathes. I can hear a cry from the earth, "I can't breathe." Among all the fellow creatures, humans have been affected the most and suffered greatly. Eventually, nuclear power will break the rhythm of the ecosystem and destroy multiple lifeforms. This should be considered an example of humanity's colonization of nonhuman nature. Nature—including human nature, especially the poor—has been subalternized since Genesis, especially since the industrial revolution two centuries ago. In *A Postcolonial Theology of Life*, I declare, "Nature is a new subaltern."[16] Elaborating on the Spivakian notion of silencing of the subaltern, I can say that, "nature is mute." This is a perfect example of subalternization of nature by human domination. To put it in a nutshell, it is not *salim* but *jugim*.

Teum (hybridizing): Energy Resources

> The world has potential for 17 terra watt nuclear energy, 700 terra watt wind energy and 86,000 terra watt of solar energy. Alternatives to nuclear energy are a thousand times more abundant and a million times less risky. To push nuclear plants after Fukushima is pure insanity. . . . It was "madness" for Japan to build the Fukushima nuclear plant in a seismic zone.
>
> —Vandana Shiva, *Making Peace with the Earth*[17]

Salim energy does not necessarily mean natural energy, as it does not matter whether it is natural or not. Both nuclear energy and fossil fuels are as natural as solar energy and wind energy. Nuclear energy means the energy released by controlled nuclear reactions. For example, heat and light produced by the sun and other stars through nuclear reactions are natural nuclear energy. Also, regarding fossil fuels, carbon is not the problem per se, but our increasing use of fossil carbon that was formed over millions of years. The cellulose of plants is carbon, as is humus in the soil. The results of burning fossil fuels are causing the climate-change crisis.[18] The problem is the broken harmony of biodiversity. Climate change is a result of the transition from the biodiversity of renewable carbon economies to a fossil-fuel-based, nonrenewable carbon economy. We need to find renewable energy resources other than fossil fuel and nuclear power and to practice how to use them in a more sustainable but less harmful way for the entire planet as a living organism. To this end, we should stop using nuclear energy and reduce fossil-fuel use. In order to make a transition to an age beyond oil, nuclear energy cannot be a *salim* solution since it only perpetuates the problems of ecological nonsustainability and has brought more dreadful consequences to the planet. *Salim* energy is, therefore, more sustainable and, at the same time, less harmful energy for all, not just for humans.

As the human economy has grown through industrialization, it has substituted other forms of energy for human resources. Here is the problem of transformation of energy into capital: Gayatri Chakravorty Spivak defines capitalism as neocolonialism, which always means the largely economic rather than the territorial enterprise of imperialism. As such, neocolonialism is another name for late capital imperialism.[19] Human imperialism over nature can be interpreted as the Deleuzian desire of capitalism. A French geophilospher, Gill Deleuze, writes, "In capitalism, desire and interest or desire and reason are distributed in a totally new way, a particularly abnormal way. Capital, or money has reached such a stage of delirium that there would be only one equivalent in psychiatry."[20] For Deleuze, capitalism is a rational mechanism, "schizophrenia": "Capitalism has always been and still is a remarkable desiring-machine: Flows of money, flows of the means of production, flows of human-power, flows of new markets: it's all desire in flux."[21] In this regard, climate change was already predicted or the expected result of the exploitation of nonhuman nature along with humans as part of the capitalistic code. When humans discovered natural energy such as fossil fuels, they dug in the ground as deeply as possible and consumed nearly everything. Thus, the problem is not about energy per se, but humanity's greedy mentality. Human beings consider energy as a solid matter that can be possessed like money rather than a source of life which flows, hybridizes, and transforms.

First, we have to recognize nature as *source*, not as resources. Shiva writes, "We are made of the same five elements—earth, water, fire, air, and space—that constitute the earth. The water that circulates in the biosphere circulates in our bodies; the oxygen that plants produce become our breath; the food that is produced by the soil and the sun's energy becomes our cells, our blood, our bones. Biologically and ecologically we are one with the earth."[22]

The natural (more than) five (multiple) elements of our body and the earth energy can transform into diverse forms such as solar power, hydro power, geothermal, wind power, and the like. Thus, energy is not stuck in one place. If it is, it will cause a break in harmony. The word *energy* is also used outside of physics. According to the East Asian cosmology in *I-Ching* ("The Book of Change"), the universe is constituted by five basic elements (fire, water, wood, metal, and earth). The earth is full of energy (*Tai-chi*, 太極) in which there are two flows, *yin-yang* (陰陽), circulating and recreating via the process of diffusion and convergence. Any organic bodies regulate themselves so as to maintain equilibrium via the *yin-yang* principle. Energy constantly decentralizes its power and divides into two principles, *yin* and *yang*, and runs through the five elements and more.

Today's climate change is a sign of brokenness of a natural harmony. How, then, do we find the energy to heal this brokenness? How do we balance nature, community, and technology? Eco-theologian John B. Cobb Jr depicts sustainable society as a decentralized process.[23] Applying this decentralized model to energy, it becomes hybrid renewable-energy systems (HRES), which are becoming popular for remote-area power-generation applications due to advances in renewable-energy technologies and the subsequent rise in prices of petroleum products. A hybrid energy system usually consists of two or more renewable energy sources used together to provide increased system efficiency as well as greater balance in energy supply. These systems include hybrid cars, home hybrid energy systems, biomass-wind fuel cells, photovoltaic-wind turbines, and the like.[24] Although these systems are not perfect, they are examples of decentralization of power which prevent one or a few particular types of energy resources from being exhausted.

I interpret power in terms of multiplicity and fluidity. It is an anthropocentric delusion that power (or nature) can be owned. For in our time, life itself is an industry, subject for management, subjugation, and social control. Michel Foucault refers to the modern coercion and diminishment of bodies in society as an intolerable power over life. Foucault views sovereign power as representing itself as negating, legislative, prohibitive, censoring, and homogenous while he recognizes bio-power as always productive, immanent, exercised, capillary, and resisted.[25] In this regard, I apply Foucault's "heterogeneity of power as sovereign power versus bio-power" to *salim* energy, which is not the institutionalized legal form of power but, rather, the bio-power that is the survival power of life.

Through decentralization of power, power can be hybridized and shared by multitudes to feed biodiversity. Thus, interconnectedness of power is power of hybridity. According to Homi K. Bhabha, one of the most important figures in postcolonial studies, hybridity is a sign of the presence and engagement of colonial power. "If the effect of colonial power is seen to be production of hybridization rather than the noisy command of colonialist authority or the silent repression of native traditions, then an important change of perspective occurs."[26] Bhabha adopts the concept of hybridity for the subversion of authority in colonial discourse and resistance against the dominant imperialist power of the colonizer. To Bhabha, hybridity is defined as "a problematic of colonial representation" that reverses the effects of the colonialist disavowal, so that other-denied knowledge enters upon the dominant discourse and estranges the basis of its authority.[27]

I refer to the hybridity of *salim* as spatiality of planetarity, which is *teum* (in-betweenness). *Teum*, a Korean term, signifies an intriguing hybrid space of spatio-temporality or spatiality of planetarity. It means a blank or a margin.[28] Can we connect this *teum* between nature and humans (which are already one) and create newness out of the interstitial hybridity? "Yes, we can, by the Power of Life." Shiva calls energy "*Shakti*," the primordial power of creation, the self-organizing, self-generative, self-renewing creative force of the universe in feminine form which is the source of all divine and cosmic evolution and also the source and controller of all forces and potentialities of nature. "The universe is an expression of *Shakti* and an infinite reservoir of power."[29] We have to deconstruct our ignorance of limiting energy in the categories of mechanical energy. Whitney Bauman calls this divine sacred power of life "the God of the gaps (*teum*)."[30] Also, a Korean American eco-theologian, Grace Ji-Sun Kim, recognizes energy as *Ruah*, the divine breath that connects *teum*: "Creation is done by the powers and energies of God's own Spirit which bridges the chasm between Creator and creature, the actor and the act, the master and the work."[31] Indeed, this is *salim*!

Salim (enlivening): Energy as the Source of Life

> Climate chaos, brutal economic inequality, and social disintegration are jointly pushing human communities to the brink. We can either let the processes of destruction, disintegration, and extermination continue unchallenged or we can unleash our creative energies to make systemic change and reclaim our future as a species, as part of the earth family. We can either keep sleepwalking to extinction or wake up to the potential of the planet and ourselves.
>
> —Vandana Shiva, *Soil Not Oil*[32]

As a way to confront the climate crisis, I, as an Asian postcolonial ecotheologian, reject both a romantic essentialism and an arrogant anthropocentrism toward nature. Overall, American ideas of the human relationship to nature as either "sublime" or the "frontier" are based on a "Disney-like fantasy."[33] Nature has been depicted as a magical place like the Wonderland of *Alice* (2010) or a reclaimed place like the Pandora of *Avatar* (2009). In both approaches, humans are alienated from nature as a third person, either a spectator or a conqueror. One is too naïve while the other is too arrogant. This denial does not close the gap (*teum*) between humans and nature. We humans are a part of this interwoven web of life. Now is the time for *metanoia.* We have to confess that the climate crisis is a humanmade problem. Then, it's our responsibility to constrain ourselves through ethical action. Indeed, climate change is an irresistible moral storm. It is our planetary tragedy. It is an intergenerational crisis. It's our past, present, and future. Nevertheless, what humans have been doing toward our planet is suicidal. Land, air, water, trees, birds, fish, rabbits, and humans—we all are vulnerable to climate change. The bottom of the anthropocentric hierarchy, the marginalized, the poor, the less powerful creatures, soil, air, water, and the like have suffered the most from climate change. Humans have colonized other planetary beings and the planet herself. Once again I announce that nature has been subalternized. It is at this juncture that we must reevaluate our ethical consciousness, as Willis Jenkins writes: "A major reason that economic activists impoverish earth is because none of [their] actions pays the true costs of the ecological resources and services it uses. Social behavior does not adequately respond to biological impoverishment because the human economy fails to value the biodiversity it uses. The inevitable consequence is overuse and encroachment to the point of collapse."[34] By the same token, Larry Rasmussen calls for our ethical framework to evolve not by leaving the cultures we know, but by expanding their capacity to be inclusive and creative. Shifting our economic activity toward growth that is both ecologically sustainable and regenerative, the difficult task of maintaining a utilitarian sense of benefit that is as broad as it is deep actually becomes reality.[35]

> You were taught, with regard to your former way of life, to put off your old self, which is being corrupted by its deceitful desires; to be made new in the attitude of your minds; and to put on the new self, created to be like God in true righteousness and holiness. (Eph. 4:22-24)

We, as the anthropocentric collective, need to cast out demonic delusions of greedy ownership and put on a planetary consciousness, which I call *salim* consciousness. Although, as I have observed, the government of South Korea is apparently eco-unfriendly, as two major *jugim* projects, the Grand Canal Project (since 2008)[36] and Jeju Naval Base Project (since 2011) show,[37] the Korean *salim* movement is carried out by many Korean women who, in the care of their homes, gather milk packs, make recycled soaps, avoid plastic bags, and engage other socially responsible tasks. The *salim* movement in Korea is not only an ecological movement but has also became an economic (recovery) movement since 1997 and the IMF crisis.[38] *Ah-na-ba-da* means saving (*"ah"kyuseoda*), sharing (*"na"nuoseoda*), exchanging (*"ba"kuauseoda*), and reusing (*"da"siseoda*). The *salim* movement functions as an ecological movement for saving the economy, and connecting the economy and ecology in terms of *oikos,* our living organism, the household.

A Korean eco-feminist, Chung Hyun-Kyung, refers to Korean women who participate in the *salim* movement as "*salimists*":

> A *salimist* makes things alive, especially dying things like the earth. 2) A *salimist* touches everything like a magician, a revolutionary to smile and grow. 3) A *salimist* recycles everything: papers, milk, cartons, glass bottles, politicians and leadership positions, gods, and life. 4) A *salimist* is an activist who

> promotes peace, reconciliation and harmony. 5) A *salimist* loves women, nature, earth, and goddess, everything feminine and feminist. 6) If womanist is to feminist as purple, *salimist* is to feminist as dark green.[39]

According to Chung's "*Salimist* Manifesto," her audience is apparently Korean women who are dealing with issues for the household and daily life. For Chung, *salimist* is actually a word for a Korean eco-feminist woman. She writes, "*Salimist* is a woman warrior who protects and enlivens the earth. She is a goddess within herself who loves and respects herself."[40] Chung's identification of *salimist* with Korean eco-feminists creates difficulty for men to engage in *salim* or *salim* movements, as it implies the work of women only. *Salimist* is not a Korean word such as violin-ist and woman-ist. The indigenous Korean term *salim* has become anglicized as well as gender-essentialized. Thus, I suggest using the indigenous Korean term, *salim-kun* (*anyone* who engage in *salim*), which does not exclude men. *Salim* does not have to be a gender-specific word.

Going beyond the *salimist* essentialist perspective, I recognize a folding and unfolding of repentance (*metanoia*) and responsibility (ethics) for taking care of and responding to the planet with humans' planetary love. *Salim* is a constant renewal, a recycling of good. Life as *salim* tends to decolonize imperial power and, moreover, pursues evolution in terms of quality, as Alfred North Whitehead maintained that all life is driven by a threefold urge "to live, to live well, and to live better."[41] In a certain sense, everything in this world at its molecular level is producing and reproducing. Thus, the enlivening spirit cannot be essentialized or genderized. Beyond the dichotomy of male and female, human and nonhuman, everything is endlessly becoming new. We must recognize that *salim* energy exists not only in women and men but also in every living and nonliving aspect of creation. All beings are *becoming together* at every second symbiotically. *Salim*, the encompassing sacred energy, is embodied in all creation. Through the divine power of life everything lives a way of life, transforming, managing, deciding, and creating spontaneously through hydridizing. This unleashing energy is the source of life, not resources. Facing the climate crisis, we have to declare a sacred "No!" to *jugim* and a sacred "Yes!" to *salim*.[42]

CHAPTER 14

Rice Is Heaven

Revisiting and Transforming Lost Korean Values in the Era of Climate Change

Meehyun Chung

Christianity, Korea, and the Environment

A long time ago when I lived in a rented house, one day I became annoyed with the ants. I immediately asked my landlady for chemicals to kill them. She replied that because they were a Buddhist family she could not kill living beings. I felt ashamed. I just wanted to have a quick solution to get rid of a common, everyday pest, but here I, a Christian, was being reminded of the value of every living being. Do Christians have less awareness of nature and the environment as being our neighbor, compared to Buddhists? After all, Christianity brought industrialization, which contributed to the destruction of nature. It is not easy to answer these questions, because the reasons for current disasters are very complex and multilayered. Nevertheless, we cannot deny a certain degree of influence by Christianity in this process of environmental disaster.

Environmentally destructive ideas among Christians are connected to a misinterpretation of the creation story in Genesis 1:28. The key is how Christian communities interpret the concept of *kabash* (meaning "to conquer"). When *kabash* is interpreted literally as a violent overcoming instead of cultivating or subduing, it leads usually to imperialistic ideas and attitudes. This kind of misunderstanding has negatively affected Christianity for a long time. Even in Korea, where we have a short history of Christianity, there remains a lack of awareness regarding environmental issues to a certain degree, as well as an attitude that human beings can make all nature an object for human purposes.

Since the Enlightenment, Christianity has contributed to the destruction of nature, which is a cause of current ecological crises. Whether or not one agrees fully with Lynn White Jr., who has strongly criticized Christian theology as the main cause of the environmental disaster,[1] it is not easy to deny this kind of complicity. White underlines the agricultural revolution and technological developments of the early Middle Ages as that which led to a change of status of humans with regard to nature: "Once man had been part of nature; now he became her exploiter."[2]

Given this background, this essay investigates the current phenomena of Korea in regard to environmental issues and explores small-scale changes in order to seek alternative solutions. Its purpose is to revisit Korean traditional values and the meaning of rice, as well as the contemporary challenges and yearning for transformative ways to have an impact on our Christian thought and behaviour.

Facing Climate Change in Korea

Regardless of religion, much of Korean culture demonstrates a reverence for nature. For instance, the Korean national anthem sings of the beauty of Korean nature and expresses a kind of prayer for God's protection. In this anthem, what is praised especially is the quality of the East Sea's waters, Mountain Baekdu, and autumn sky.

> Until the day when the East Sea's waters are dry and
> Mountain Baekdu worn away,
> May Korea prosper forever
> God be at our side

. . . Autumn sky is void and vast,
high and cloudless,
the bright moon is our heart,
undivided and true.
. . . With this spirit and this mind,
Let us love our beloved land and give all loyalty,
in suffering or in joy.

Before Christianity was systematically introduced to Korea through North American missionaries, there was already in people's minds and faith a creator. There was also a deep respect for and forms of prayer to a heavenly being. The concept of creator and protector has existed throughout Korean history. After the Bible was introduced in Korea the perception of a redeemer was first introduced as necessary to spirituality. However, this understanding of salvation was limited—one-sided and one-dimensional. As a result, it was understood simply as personal salvation without a dimension of otherness, including nature. The strong interest in personal salvation is an obstacle to open minds toward social issues because there is a lack of awareness of an interconnectedness to others and a notion of communal goodness.

Due to its geographic location, Korea's winter is affected by continental winds, while its summer is affected by the Pacific's winds, temperature, and humidity. Korean culture and customs are closely related to this weather and the seasons. We used to say in Korea that we have four distinct seasons, as well as a characteristic cycle of three cold days and four warm days. However, this expression is changing now due to climate change. There are no more significant characteristics to distinguish among the seasons. Summer has become more humid and hot while winter has become colder. As a result, we spend more energy for air conditioning and warming instead of regulating temperature through properly protected architectural forms. Evil never ceases to circulate. Without using drastic words, we note the distortion of creation in the prevailing realities in every place of our globe even if it is affected differently in the current situation. The question is, Do we recognize the urgency of this problem and are we ready to change our way of life drastically? Additionally, due to division of the lands between North Korea and South Korea, ideological conflicts remain instead of spreading a more universally minded understanding of salvation. Dualistic divisions and thought between friend and enemy is a great hindrance to building up a constructive correction to our entire way of life. Insecure conditions through division create more pressure generally to manage life by a more individualistic perspective and drives competition among people with egocentric ideas. Pro- and anticommunist debates still remain as major issues even within the churches. Therefore, climate change as a result of global warming has not come up as a burning issue in society or in church.

Current Tendencies and Challenges

In the traditional understanding of nature, the Korean notion of Spirit used to be deeply engraved in the culture, history, and religion, but it has changed in the present day.[3] So, it is not just an idealized, uncritical traditional value. The important thing is how it is implemented as a daily praxis. It is all about everyday reality. But what is the current reality? I would like to illustrate the current situation in Korea from three aspects.

Godlessness in our age

Since the liberation of the country and the Korean War, the Korean church has grown numerically very quickly. Korean Protestantism in general was due to the very North American-oriented influence of Christian missions. Without having much awareness of the indigenization of Christianity in Korea, Americanism has heavily influenced the development of the Korean church. The first numerical growth took place in the beginning of the 20th century. The second stage of expansion was observed in the 1950s and '60s, where general national economic growth was visible and booming, because everything at this time had to do with the rapid growth and recovery after the wars. American Christianity was considered a symbol of material blessings. The core message of the gospel is therefore confused with American Christian culture. In the 1970s and '80s the democratization process was important during the time of dictatorship.

Power mechanisms and the monopolization of power were criticized. Nevertheless, the awareness of the oppression of nature by humans and the necessary attention to the contemporary world power balance had not yet emerged. Material growth in this period was much more important than anything else. Ecological awareness emerged only later in the 1990s. Both economic growth that lacked solidarity with marginalized people under neoliberalism as well as the trap of globalization gained more attention because the disparity between rich and poor became a more serious issue.

Meanwhile, once Korea reached a certain standard of living according to the UN Organization for Economic Cooperation and Development criteria, the ecclesiastical landscape also changed slowly. According to census surveys, the rate of religious affiliation has decreased, that of Protestants the most, from 19.7 percent to 18.3 percent, or by about 1.4 percent in the last ten years.[4] The recent trends show an ongoing decrease. Generally, they show that since many Koreans have become rich, even though economic disparity is more visible, religious interest has decreased. This means that religious matters are mostly used to fulfill material needs; the relationship with God is used for material prosperity.

Nature oblivion

In this contemporary digital age, it is all about image. The admiration of real nature is often forgotten. Nature is encountered instead in images on a computer screen or smartphone. Conversation with nature happens in this digital way. The possibility of having an intimate relationship with nature is rather limited for those with a lifestyle grounded in the big cities like Seoul. Moving to the cities and the resulting urban/rural divide is also a big problem. Because of weak competitiveness and a lack of access to public facilities in the countryside, rural areas often do not have enough resources for survival. As a result, Korea has developed as a major city-centered nation. The urban poor suffer from not having proper living conditions, while elderly people remain in the countryside. Due to urbanization and resettlement in the big cities, there are few green spaces and no special opportunities for citizens to have a proper relationship with nature. Instead of nature, ongoing construction takes over everywhere.

Ignorance of others

The new means of digital communication, such as Facebook, smartphones, Google, Twitter, and so forth, allow versatile communication with people from a distance, but they also prevent intimate, face-to-face conversations with others. The profit-oriented economic structures and the prevailing mindset are pronounced everywhere. Competition is felt strongly in this elbowing society. Busy city life forces people to ignore neighbors. Additionally, because of the dominant "pure blood mindset," it is hard to accept other people from outside of the country. "Mixed blood" is disrespected. And, based on Orientalism, there is conscious and unconscious racism in society as well, and people are treated differently on the basis of skin color.

Traces of militarism

This kind of threefold forgetting and ignorance of God, human beings, and nature leads to destruction on different levels. There are various cases which show the great lack of awareness of the environment as a co-creature.[5] Korea used to be called a land of beautiful scenery, but many mountains and rivers have been destroyed for the benefit of leisure industries like golf and skiing. Let me mention two major examples here. First is the four rivers restoration project. According to the Korean government, the four major rivers restoration project was launched as a multilayered green growth project in 2009. The four rivers refer to the Han, Nakdong, Yeongsan, and Geum rivers. However, in 2014 it had already been demonstrated with scientific evidence[6] that, rather than restoring nature, it has been destroyed. Second is the Miryang project for high-voltage transmission lines. This is a project to build high-voltage power lines connected to a nuclear power plant. Nuclear energy has not been abandoned because it is ostensibly economically cheaper. Profit-oriented thoughts always come first.

But a major destructive problem has been caused by the Korean War, which left huge traces of tragedy and suffering throughout Korea. Through the division of Korea, many kinds of suffering and social problems in both the personal and the public areas have occurred. Due to Korea's geopolitical context and complexity, this has been frequently used for the militarization of Asia Pacific in the past and present, and is likely to be so in the future. United States militarism was responsible for the pollution of many rivers and streams. Since 2000 Korean society has become aware of the way in which U.S. soldiers dumped toxic chemicals into the main rivers of Korea. Due to the unjust Status of Forces Agreement between Korea and the USA (SOFA 3), this kind of water pollution continues.[7] Relocation of American military in Pyungtac has resulted in huge tracts of Korean farmland being no longer workable for farming. This has both destroyed farming communities and inflicted extensive environmental damage.[8]

On the island of Jeju, a large U.S. naval base will continue to be built. One of the most important incidents is related to Gangjeong Village on Jeju Island, which is a beautiful volcanic island and a well-known place for honeymoons due to its natural beauty and location. But it has also been used for exile due to separation from the peninsula and the historic difficulty of transportation. In modern history it has also been used for ideological conflicts, like the Jeju 4.3 Massacre. Since publicizing the construction of a naval base on Jeju in

2002, the current ongoing conflict emerged.[9] It is not only a conflict between construction and destruction, protection of environment and economic interests, but it is also based on ideological dualism. Korea is still living under this kind of spiritual destruction of separation and dualistic ideas and trauma of the war.[10] Of course, the pollution of water or environment in the Korean peninsula is caused not only by U.S. militarism, but also because of many other related reasons regarding the division of Korea. Due to unstable living conditions and psychological pressure, people are generally more oriented to achieve something quickly, instead of taking time for sustainability. This kind of speed-oriented lifestyle brought us, positively, quick development and industrialization. But now we are gradually aware of the dark side of this kind of material-oriented, one-dimensional development.

Threefold Thought and the Meaning of Rice in Korea[11]

Threefold thought in Korea

In Korean traditional thought there are the principle and concept of a triple relationship of "heaven–earth–human." It is observed in order to describe God's being and what it means to have a relationship with God, human being, and nature. According to Korean myth, a triune god-being, Hwanin, Hwanung, and Dangun, made heaven, earth, and humankind. From this myth the threefold thought and principle derives. Therefo re, the number three has been recognized as a perfect number. This "heaven–earth–human" idea has affected Korean culture and the Korean language broadly, and has been adopted deeply by Korean religions. It is used dialectically and universally. This perception is related to creation and to giving and sustaining life.

This triune principle is not only a matter of Asian terminology.[12] It is a central principle of the Christian message.[13] This threefold relationship is established with God, with our fellow creatures, and with our fellow humans.[14] The threefold thought underlines harmony and the combination of heaven, earth, and human beings. Additionally, Christianity highlights not only this kind of universal harmony, but also God as Creator, Sustainer, and Saviour.[15] Between God as Creator and all co-creatures (*mit-welt*) is situated the relationship of creation and covenant. To echo the famous words of Karl Barth, "Creation is the external basis of the covenant. The covenant is the internal basis of creation."[16] Compared to Christian thought, in the Korean threefold thought there is no perception of this relationship between creation and covenant through creator and creature. To conceive this universe as the house or household of God is, rather, a genuine Christian perception.

On the other hand, a universal dimension and understanding was included in Korean traditional spirit and religious thought. However, this value was lost through the processes of industrialization and Western-styled modernization, which was brought through Christian mission. Additionally, in the Korean case, patriarchal masculinity in militarism and greed for material richness without thinking of communal goodness were great obstacles to developing this valuable traditional thought. To recognize the authentic Korean thought, as well as to understand fully the Christian spirit, it is necessary to recover this threefold thought, to apply them in all dimensions of social life and to practice them day by day. This means it helps to correct human-centered ideas and, furthermore, male-centered androcentrism. The well-being of *all* parts of the universe is important and not just some parts of it.

The Greek word *oikos* refers to a house or household in a narrow sense, and to the universe in a broad sense. This Greek word root alludes to the inhabited world and includes all elements of the universe, and not only human beings. There are many verses in the Bible in which God and the *oikos*-house are connected. In Hebrews, our globe is called the house of God. In this house, Jesus serves as high priest. We live in this house together with other creatures as "God's family" (*familia Dei*) and as "guests."[17] We have to keep regulations in order to live on this planet. "(For every house is built by someone; but the builder of all things is God.) Now Moses was faithful in all God's house as a servant, to testify to the things that would be spoken later. Christ, however, was faithful over God's house as a son, and we are his house, if we hold firm the confidence and the pride that belongs to hope" (Heb. 3:4-6). "Therefore, my friends, since we have confidence to enter the sanctuary by the blood of Jesus, by the new and living way that he opened for us through the curtain, that is, his flesh, and since we have a high priest over the house of God" (Heb. 10:19-21). Thanks to the work of Jesus, the curtain in the sacred place has been torn apart. Through the torn curtain, the gap between the secular and the sacred is leveled. What is important is the aspect of the house of God, which has been restored by Jesus Christ. God has prepared this house and we are part of it—what Paul describes as "God's garden and God's building" (1 Cor. 3:9). This idea is similar to the Shekinah in the Kabala, or the eco-feminist ideas of Rosemary Radford Ruether, Sallie MacFague, Ivone Gebara, and others, which emphasize a panentheistic dimension in eco-awareness and interrelatedness between human beings and the universe; here the house of God is connected with the body, planet, and the universe seen as a whole. The Korean "heaven–human–earth" idea is not much different.

The meaning of rice in Korea

I want to transfer this idea of *oikos* in the Bible to the Korean context regarding the verb *build*. First, for a terminological explanation of the verb I must emphasize the importance of rice in Korea. Rice is a major staple food and keeps people alive. It has an important economic value, and its greatest importance is that it is a component of life. "Rice is heaven" is one of the best-known poems by Chiha Kim:[18]

> Rice is heaven
> Like heaven you can't possess it as your own
> Rice is supposed to be shared
> Rice is heaven
> . . .
> When rice goes into the mouth,
> You keep heaven in your body
> Rice is heaven
> Oh, rice is supposed to be shared

Rice is one of the most important elements of life. The verb that is used in connection with rice cultivation and rice cooking is also used in house building. In Korean, this verb is used equally in the different contexts of rice, clothes, and housing. This verb is used in the agricultural culture especially for joint collaboration. Furthermore, it is used for sustaining life, as in wearing, eating, and staying. That is, it is not simply a matter of creating, but the result is on behalf of the community's goodness. A building or house is not merely an object to be constructed; staying alive is also an important aspect. The house is, in this sense, not an object for investment, and not merely an object of development. It remains as a community and, thereby, for the well-being of we-ness. God makes a human being a co-worker or fellow worker (1 Cor. 3: 5-9). It is an invitation to keep constructively life and not death.

Rice has significant meaning for life in Korea in general. In 1441 King Sejong invented the first device in the world to measure rain. This rain gauge included the combination of "heaven–earth–human" because it was used primarily for rice cultivation. Rice is a result of these threefold components: the creator gives rain, the soil produces rice, and people's work combines these facets. Since Korea used to be very dependent on agriculture, it is important to wait for the rain. Rain is one of the key success factors for the cultivation of rice, because rice is a water plant. In the Korean language, the polite form is to be observed in describing the coming of rain. It implies the importance of rain. To receive rain is an act of gratitude.

Since rice is a water plant, water is the most important component for rice.[19] In order to harvest one kilogram of rice, two tons of water needs to be provided. Therefore, rice and water are closely connected. Rice is a very valuable product. "It could be dedicated to God"[20] in Asian culture. Most Asian folk religions have something to do with rice. This shows the importance of rice in this region. Asia includes and embraces old traditional religious values, but there is no unifying commonality of culture, religion, and language. Nevertheless, rice could be said to be one common denominator. Rice provides the basis for cultural and religious values as well. Korea has a very appropriate natural climate for rice planting. Rice needs hot weather and humidity such as is found in the monsoon season.

Rice, therefore, is assimilated to the natural conditions found in Asia. On the contrary, in the USA, Australia, and other countries, rice is artificially developed with mechanization and chemicals. It is, therefore, antienvironmental rather than eco-friendly.[21] Filtering and the preservation of water by rice planting prevent floods, landslides, and erosion. The air and water are filtered through the rice-growing process. Rice farming is a natural preventative against erosion in the Korean peninsula.[22] Rice planting, as an environmentally friendly process, is one of the most important examples of sustainable development in the 21st century. Furthermore, traditionally for rice cultivation, special cooperation of labour, known as *Durae*, has been used. *Durae* was a farmers' co-operative group that developed around the 18th century and disappeared around 1950. But this method originated around the 4th century in the Korean peninsula in the Samkuk period. All of the steps in the rice-farming process—nursery, transfer, irrigation, harvest, and so forth—were organized by the farmers' cooperative. Thus, it was a collaborative method of community work. A small farming system was protected through this community-oriented method. *Durae* was a civil movement that was organized according to the season and the labor cycle. It was carried on in a spirit of mutual help. *Durae* was based on independence, voluntary responsibility, interdependence, and solidarity. The farming work was accompanied by music and cultural festivity, which was one of *Durae*'s most important components. In our age of globalization, we need this win-win method of interdependence. *Durae* provides a good example of it, because everybody is connected to the communal work and the harvest and thus work in a spirit of justice and cooperation rather than self-oriented egoism. The method of *Durae* implies strong consideration of otherness, especially physically weak people and people on the margins.

The whole process of rice farming is in harmony with nature. During the Chosun dynasty, in the early 19th century, Hakyou Chung composed poetry about rice farming. His poetry reveals the custom of daily rice farming, and encourages a rhythm of labor which is in harmony with the seasons. Rice production is in harmony with the sun, water, and the life of the seeds, earth, and the labor of human beings. Rice farming is adapted to nature. It makes us aware of the interconnectedness between nature and human beings, thankful to God as Creator, and respectful of the wisdom of our ancestors. We tend to measure everything according to speed and technology. But even if the process is very slow, we need to relearn the rhythm of the seasons. A community-oriented, collaborative, and eco-friendly outlook in the spirit of *Durae*, an emphasis on play and folk songs, and the strength of traditional religion could be emphasized and restored through Korean feminist theology.

Rice farming is called the basic ground of heaven and earth, but the farmers who engaged in this holy farming didn't receive proper compensation. They became the victims of industrialization, urbanization, and the free-market system of the World Trade Organization (WTO). Due to this kind of urbanization, the major cities in Korea became overcrowded while the countryside became empty. This caused many social problems. Our agricultural system broke down with the introduction of the new liberal trade policy. Losing the meaning of the agricultural way of life caused many kinds of problems in the health system, including that related to climate change. Endless expansion without thinking of sustainability caused this problem. Korea is a small country and there is little opportunity for price competition, so the protection of agriculture depends on the attitude of the consumer. Government policy can help reduce the cost of rice production and enable young people to remain as farmers on the land. On the personal level, the most important thing is awareness on the part of the consumer. Korean rice production is more expensive compared to the cheap production in China, the USA, and Australia. If Korean rice production is to prosper, consumers must stand in solidarity with Korean farmers. Consumer awareness and solidarity between producer and consumer[23] can provide important support for sustainable agricultural development in the Korean peninsula. To keep a certain amount of this kind of agricultural product and real recognition of producers will bring constructive solution against climate change, because it lets us correct our greed-oriented capitalistic expansion.[24] An eco-friendly agricultural method could also contribute to environmentally friendly development in general.

Transformative Paradigm Change and Responsibility of Korean Churches

There is a unique potentiality around the Demilitarized Zone (DMZ)[25] to develop in a totally different way even though there is the issue of landmines which threaten the lives of many people in both Koreas.[26] This kind of human-made device already has produced countless victims in Korea and a tragic accident can lead to the ongoing suffering of a whole family.

One positive side regarding the protection of biodiversity in this nobody's land in the Korean peninsula can be mentioned. Due to the lack of human destruction, many species remain alive here. In spite of this well-known biodiversity in the DMZ, South Korea's efforts to nominate the DMZ for the Man and the Biosphere Programme (MAB) of UNESCO has been deferred due to a lack of consensus between the two Koreas. The entire Korean peninsula should become a Demilitarized Zone, not just partly around the military demarcation line. This means that the ceasefire or armistice agreement should be changed into a permanent peace agreement, which could contribute to healing the people who are affected by long-sustained fears, threats, and hostility by this division of the Koreas.

Countless people have already suffered from the Korean War and division of Korea. To this day, all young male Koreans, with minor exceptions, spend precious time in military service. Ongoing tension and conflict happens continuously. Therefore, militarism influences not only the military system, but the entire Korean culture. The whole society is affected by this patriarchal military order-obedience mechanism. Those who possess hegemonic power dictate everything instead of promoting the spirit of empowerment and embracing diversity. A very monotonic, linear idea dominates and does not consider otherness and complexity. This kind of power chain is hard to break because it is easy to monopolize power. Good governance is not a question at all. The ruling government's only solution is to keep this system. It is a great hindrance not just to the development of democratic creativity and the acceptance of diversity, but also to reaching a consensus by recognizing different ideas and thoughts.

The speed-oriented psychological pressure for economic growth, the lack of a communal oriented mindset and ecological awareness, and other factors are also influenced by this uncertainty of division and potentiality of war. We need to be liberated from all kinds of fear and mechanisms of oppression that are related to the issue of national security. Breaking the chain of all bondage to fear is more necessary than ever before. Where nature is destroyed largely due to human economic interest, there is a new opportunity for Korea. Someday a peace treaty should be completed for peace in Korea. Afterward, this zone of the tragic Korean Demilitarized Zone (DMZ) could be transformed into a new land, a "Dream-Making Zone (DMZ)." But this is not just an eschatological utopian dimension. "The LORD All-Powerful will destroy the power of death and wipe away all tears" (Isa. 25:8). In the process of marching toward God's reign we can reach our human goal through protection and preservation instead of human destruction. The DMZ should not be a symbol of the Cold War anymore, but a symbol of reconciliation through the threefold relationship of God, human beings, and co-creatures.

The restoration of ancient wisdom that is related to femininity is generally important toward sustainable development.[27] For us in Korea it means the restoration of harmony between nature and human production in agriculture and the spirit of *Durae*. Accommodating the community-oriented mind, agricultural heritage, and tradition as positive religious values should be reinvestigated and promoted again, because, due to quick industrialization and uncritical Westernization, this kind of value is gradually dismissed. Korean cultural values, however, should not be idealized, but critically evaluated.[28]

Like liberation theology,[29] Korean Minjung theology was not particularly focused on the environmental issue until recently. This spirit, however, is part of our cultural heritage and since 1990 it has become an important item on the agenda of Korean theology and in the church movement "Life-*Salim*." Indeed, many kinds of life movements are now flourishing in Korea in general, but also in churches in particular. Liberal progressive Christianity has declined and the Christian world has become more conservative. To raise a prophetic voice in a neoliberal global era is related to a paradigm shift from blood-oriented, individualistic "red salvation" to "green salvation" that includes all creatures. To underline Jesus' blood as a condition for salvation is important. Salvation is ultimate healing not only of the human being, but also of all creatures. Due to androcentric ideology, it is not easy to accept this kind of broad dimension of salvation. God did not send the Son for the salvation of humanity alone or give us a partial salvation. Rather, the gospel is the good news for every part of creation and every aspect of our life and society.

"Mission from the Margin" was the title of the new mission document adopted at the 10th general assembly of the World Council of Churches.[30] Indeed, mission from the margin has not been recognized fully since Constantine's new era of Christian history.[31] Christianity was recognized by the Roman Empire, but it no longer possesses the essential vitality that was more possible when Christianity was a minority and persecuted religion. We need to restore listening to the voices of those at the margins who suffer firstly and vastly through climate change. It is therefore vital to recognize God's mission in a cosmic sense and to affirm all life, the whole *oikoumene*, as being interconnected in "God's web of life."[32] Facing climate change and global warming now is a pivotal moment to radically change our thoughts and attitudes about living, especially among Christians. "Humanity cannot be saved alone while the rest of the created world perishes."[33] It depends on how we overcome the ongoing secularization and capitalist-oriented world and the enslavement of the human mind. In a time of growth, where everything is measured according to physical criteria and figures, we cannot resist recognizing human limitations and emphasizing the variety of connectedness. Although the influence of religion has lost some weight, it is important to strengthen healthy religious functions and thereby give positive energy to share control. "Eco-justice cannot be separated from salvation, and salvation cannot come without a new humility that respects the needs of all life on earth."[34] "God's love does not proclaim a human salvation separate from the renewal of the whole creation. We are called to participate in God's mission beyond our human-centered goals. God's mission is to all life and we have to both acknowledge it and serve it in new ways of mission. We pray for repentance and forgiveness, but we also call for action now. Mission has creation at its heart."[35]

Conclusion

To call for actions in the here and now means to start with the small things in our daily lives, including even the use of electronic devices at home, a practice that is related to energy consumption on a larger scale. Ultimately, we need to reach for the elimination of nuclear-power stations and work to use renewable power. A material, profit-oriented mindset is short-sighted and does not lead us to think of a realistic kind of sustainability. Therefore, Christian voices should be raised to correct the appearance-oriented arrogance of human beings. In order to correct the current material-prosperity-centered tendency of Christianity it is necessary to restore this kind of holistic perception of a relationship with God, human beings, and nature. It could contribute to differentiated development against the one-dimensional capitalistic market approach and the uniformity of unsustainable development in this land of dualistic ideological division and excessive militarization. Instead of a dualistic extreme conflict between pro- and anticommunism, we need to have a dialectical vision for the co-existence of Sang-Saeng or life-giving mutuality and "symbiotic life" in order to construct a "Dream-Making Zone," to replace the old Demilitarized Zone, which is a symbol for broken relationships, hostility, and separation. Leaning on the power of the Holy Spirit we could be transformed to construct this new Dream-Making Zone as a symbol for interconnectedness, gratitude, and reciprocity. Seeking after the forgotten threefold thought in Korean tradition might be a first step to approach this goal. A third way is necessary in order to overcome the mechanical dualism of tension and dichotomy like white and black, wrong and false, capitalism and communism. Afterwards, dialectical synthesis could be derived through this tension.

It is very important to restore rhetoric about climate change and biodiversity for sustainable development in terms of the triple relationship of "heaven–earth–human." This has to do, in the first place, with recognizing the diversity of nature and our life system of interdependence and interconnectedness. It promotes diversity in every step of life and transforms culture in the private sector as well as public sphere.

The Christian spirit of resistance against the status quo of oppressive power is inherited throughout Christian history, even if it sometimes looks like dying embers. The glimmers of hope are never extinguished and have been sparked in every epoch. This revives the dream of liberation. Small-scale dreams at the micro-level could accumulate and then the large-scale dream at the macro-level could be reached. Throughout history, a creative minority has brought the spark to maintain this hope of the transformation of Christianity. Thanks to them, Christianity has remained alive even though they were frequently persecuted and ignored. Our part is to commemorate them and to make this tradition more visible.

The reduction in numbers of Christians has now also become a problem in Korea. But, rather than being afraid of losing members, I think it is important to help the majority of evangelical Christians who are biased toward growth not to regard natural disasters as God's punishment or to see many kinds of human error. It is important to raise awareness of environmental and economic justice, to jointly exercise political and social responsibility, and to carry on.

There is no denying that, through Western churches, many indigenous cultures and economic systems were destroyed. Nevertheless, today it is also important to acknowledge the positive aspects of the mission. We all face one of the biggest turnarounds. Reappraisal of the negative effect of Christianity in the church's history is ongoing. But that should not be an obstacle; the good news continues to proclaim that one can combine word and deed.

As the parable of the good Samaritan (Luke 10:25-37) illustrates, the person who showed mercy and pity was not a religious leader in the established mainline institution, but a Samaritan who was an ethnically marginalized person. Our nature has been beaten by robbers. We are those religious leaders who did not make any efforts to rescue the victim and we are also the robbers because of our material greed that prompts us to take from others, including nature. *Metanoia* and repentance are among the strengths of Christianity. *Metanoia* makes space for new things by first emptying old spaces, despite their weakened positions. The churches could contribute to this great transformation, because the emphasis of *metanoia* makes such an action possible. Great action is not just what is large and important, and small transformation is not small. Each step, large or small, is important. The emphasis on rice in this agricultural heritage, which is a common base in Asia for harmony and community, can provide an alternative to the global spread of a liberal trade policy in the 21st century. It leads us to less destruction of our globe in this era of global warming. From earth we need to learn again to give up our materialistic greed and to promote interconnectedness toward life.

CHAPTER 15

Art for Climate Justice

Lucy D'Souza-Krone

This chapter will examine the care and concern for the environment expressed through the Bible by means of painting. A series of paintings was exhibited at the 10th general assembly of the World Council of Churches in Busan, South Korea, in autumn 2013.

The main intention of this series of my paintings is the preservation of creation. They refer to biblical texts—here interpreted within the context of our present age by means of art—in order to point out how human actions influence the future conditions of life on this planet.

The paintings are intended to contribute to a new vision of creation and to stimulate a rethinking of our way of living. To achieve the much-needed fundamental change of lifestyle and forms of economy, we need a spiritual change, too. The paintings aim to contribute to a new understanding of the world as God's creation. I would like to exemplify this through four of my paintings.

God's Eye Watches over the Earth

> The land, which the LORD swore to give . . . that is . . . watered by the rain from heaven. Yahweh your God takes care of this land, the eyes of the LORD your God are on it always. . . . If you shall faithfully obey my commandments . . . I will give your land rain in season, autumn rain and spring, so that you may harvest your corn, your wine, your oil. I shall provide grass in the fields for your cattle, and you will eat and have all you want. (Deut. 11:9-15)

All this is visible in this painting. The eye of God in the center watches over the seasons: winter, spring, summer, and autumn. The blue colours stand for the cold seasons, gray for the snow (below right), the warm colours for the summer (upper left) and autumn (upper right). When God's eye watches over the earth, everything goes well for humans and all living creatures.

But climate change influences the cycle of weather and endangers the seasons' cycles for farming. The painting aims to sharpen our consciousness to respect and preserve the blissful, "bless-full" rhythms of nature and the ecological balance—so that we do not destroy this wonderful earth.

Tree of Curse and Blessing

This painting puts both possibilities—blessing and curse—dramatically in front of us.

> If only you obey the voice of Yahweh your God . . . blessed will be the produce of your soil. (Deut. 28:2-4; see also Lev. 26)

> Then there will be water and blue sky, wine, oil and bread to strengthen us and to delight our hearts. But if we do not listen to the voice of God, the earth will become a desert by reason of its inhabitants, in return for what they have done. (Mic. 7:13; see also Jer. 9:11-13)

Both possibilities are symbolized in this painting by the fruitful and the dried-up sides of the tree. Our behaviour today decides if we are to have curse or blessing on this planet—whether the dried-up side of the tree is to become a reality for many regions of the world. That challenges us to think about what kind of future we are heading for. There is always the chance for change, as by the grace of God we can find the right way.

The Sun as Motor of Life

This painting guides us to choose the right way.

The sun is the motor of earth's energy circulations. The small circle in dark blue stands for the oceans whose upper waters get warmed up by the sun and evaporate. Wind arises out of the difference in temperature and the pressure of warm and cold air, which can be used as energy. The clouds, driven by wind, rain down, filling rivulets and streams, and from that water we can gain energy. Water and sunlight cause vegetation to grow and biomass to emerge.

We humans can make use of the sun in many ways—for example, by cooking with solar cookers and even more by using solar collectors—they form the big blue circle. This painting invites us to use renewable energies on a large scale. Heaven provides us with energy in abundance, with the sun as motor of life.

Do you already celebrate the Harvesting of Renewable Energy? "Celebrating the Harvesting of Renewable Energy" can be performed all over the world as a common feast by everyone on earth—as the whole world receives renewable energy. It could connect cultures and religions worldwide and would be a contribution toward a "culture of climate," which has to be developed in this century.

The River of Life

The last painting in this series deals with the final destination of all life.

> Out of the heart of the holy person springs the river and in her presence everything thrives. Along the river, on either bank, 24 trees bring forth all kind of fruits all year through and their leaves serve for medicine. (Ezek 47:1-12; Rev. 22:1-2)

The fishing nets are a sign that fish will be "plentiful" (Ezek. 47:10). In total, you see 153 fish in the painting, as in the Gospel; in ancient times this was a number for fullness, which can be understood nowadays as a symbol for biodiversity, too. The river is comparable to a heart full of love, that the people around us are helped to grow in the stream of life; we are to bring healing around us.

The sun and the moon stand for day and night, and so the river flows constantly. May you as a person be transformed more and more into such a river of life to preserve the beauty of nature.

I made big poster prints of each painting for the exhibition at the assembly in Korea. With these prints I was able to show the images and, in the future, in different places in the world, they will inspire more people.

In addition to the Busan exhibition I presented a workshop at the Catholic Church day in Germany in 2014, entitled "Discover and Dance Your Element." Participants were asked to choose their favorite element, to paint a mask accordingly, and, finally, to present their chosen element by dancing with the painted mask to the rhythm of music. This activity helps to create awareness to care for creation.

The Catholic Church in India has chosen my painting "The Ecological Angel" for an India Environmental Justice Sunday and added the words: "The planet earth needs angels of conservation! The fragile earth needs agents of environmental justice! You and I can respond to this need of the hour."

That is exactly what my series of paintings, "Art for Climate Justice," aims for.

Part Three

Interfaith Initiatives and Care for the Earth

CHAPTER 16

Divest and Reinvest Now!

Faith Communities, Fossil-Fuel Divestment, and a Clean-Energy Future

Fletcher Harper

Is it immoral for religious groups to invest in fossil-fuel companies? Is there a moral imperative for these groups to invest in a clean-energy future? In a word, yes. This chapter will explain why.

Since 2012, divestment campaigns have sprung up at hundreds of universities and religious institutions internationally, in dozens of U.S. municipalities, and in a number of environmental organizations. To a lesser degree, a number of these institutions are considering reinvesting in a clean-energy future.

Initial Reactions

For some, divestment is galvanizing and, for others, "divestment fatigue" is the response. In recent years, many religious groups have debated divesting from companies working in Sudan. The Presbyterian Church USA's Israel divestment debate has created bitter controversy,[1] and, for some, another divestment campaign is just too much.

Others oppose divestment. Shareholder advocates argue that it decreases the impact religious groups have on industry by eliminating the chance to engage management. Investment managers worry about lower returns and greater risks.[2] Others argue that divestment is the wrong tactic in the fight against climate change, citing government policy, not corporations, as the proper target and noting that fossil-fuel companies also invest in renewable-energy projects. Still others suggest that fossil-fuel companies are too big to be hurt by divestment.

These are arguments that all must engage when entering the fossil-fuel divestment debate.

The Current Context

Divestment and ethical investing debates happen within specific contexts. Four aspects of our current context are most relevant:

- Accelerating global greenhouse-gas emissions represent a grave crisis to which political systems have been unable to respond.
- The developing world's growing need for clean energy represents a moral imperative.
- We have the ability to increase the growth of renewable energy more than ever before.
- The lack of a convincing, overarching narrative for solving the climate crisis argues in favor of an approach such as divestment and reinvestment.

Risk: Accelerating emissions, unresponsive political systems

Studies agree universally: the world's greenhouse-gas emissions are continuing to rise rapidly. The human community has never faced such certain and devastating consequences if action is not taken. Apart from nuclear war, the world wars, and ancient plagues, climate change dwarfs previous threats to humanity in its scope and severity. The magnitude and immediacy of the threat in itself represents a powerful argument in favor of divestment, an act which inherently signals the need for dramatic change.

Yet, in the face of this hard data, a critical mass of society's leadership has not yet seen fit to force a large-scale legislative, economic, and cultural response to the climate crisis. This failure by political and economic elites is as great a problem as the rising emissions themselves. Such a long-delayed response represents another indication that nonviolent, disruptive action like divestment is necessary. What else will wake us up?

The clean-energy poverty of developing nations

A second aspect of our current context is the developing world's need for more, cleaner energy. For the extremely poor, energy is life saving. As energy becomes more available, infant mortality drops and life expectancy rises.[3] As billions of people emerge from crippling poverty, they need more energy. Developing countries also need cleaner energy. Air pollution is one of the largest causes of death in the developing world, due to the use of dirty fuels in cooking stoves and coal plants which create large public-health hazards and internal-combustion engines that generate harmful emissions.[4]

We need a world with much more clean energy than we currently have. A no-growth energy strategy is neither possible nor acceptable for the billions afflicted by hunger, material deprivation, and energy poverty. A divest-and-reinvest strategy is one way for faith groups to signal forcibly that the time for a strong commitment to a clean energy future has arrived.

Ready for scale

A third element of our context is that a growing number of technologies, policy tools, and social-science-based approaches to reducing greenhouse-gas emissions are poised to grow to a far larger scale. The wind-power industry has grown at a rate of 25 percent annually during the past five years, and the solar industry by a 50 percent annual rate.[5] And yet, the world's governments spend approximately USD $2 billion per day to subsidize fossil-fuel use,[6] more than six times the subsidies enjoyed by the renewable sector.[7] It's long been recognized that shifting these subsidies would greatly accelerate the rise of a clean-energy future.

Narrative free fall

From the late 1980s through 2010, a unified narrative framed our understanding of how the climate crisis would be solved. Education and grassroots organizing would raise awareness, which would lead to legislation, regulation, and financial incentives to address climate change at the domestic level. A U.S. commitment to domestic action would lead to an international climate-change agreement.

It didn't work out that way, and given the present makeup of its House of Representatives, there is little chance that U.S. climate legislation will be passed in the coming several years. This means that we are proceeding toward certain danger without a plan or a plausible narrative to guide us. Ironically, this former narrative has disintegrated just as more people are experiencing climate-change-related impacts such as more intense storms, droughts, or wildfires.

We have entered a "dark interval,"[8] a time in which a secure sense of reality (a stable climate) and source of hope (the prior narrative) have died. Dark intervals are times of insecurity, fear, and despair. They are also times when new possibilities can emerge in the space created by these challenging, dangerous circumstances. Our current "dark interval" cries out for a major shift in values, policies, and practices around energy. Past energy transitions—from one dominant source of energy to another—have taken five to ten decades or more.[9] The science is unequivocal; we need faster action to avoid massive suffering.

To create a tipping point for large-scale action on climate issues, a firm expression of resolve is required. Divestment and reinvestment are a uniquely effective way to signal such resolve.

The Biblical Rationale

Those supporting divestment have strong biblical arguments to support their position. Here is a brief summary of the three most important arguments.

Creation's goodness, God's covenant with life

In Genesis 1, God repeatedly calls the various aspects and elements of creation "good," and at the conclusion of the sixth day calls everything that has been created "very good." Biblically, creation's goodness and value to God are indisputable. Further, in Genesis 9, God makes a covenant—a binding commitment—not only with all humanity but with every living creature, promising that God will never again destroy as was done in the flood. This demonstrates a commitment to all forms of life, now and in the future.

Climate change's massively destructive impact on creation is described elsewhere in this book. Clearly, such human-caused climate change is counter to God's will. The primary reason that we have no international climate treaty is because of the fossil-fuel industry's obdurate opposition to such a treaty. This industry, for example, spends over USD $400,000 per day to lobby the U.S. Congress[10] and over USD $1.5 billion per day searching for new sources of fossil fuels, far beyond the USD $244 billion invested globally in renewable energy.[11] Given the scale of harm created by climate change, how can such industry action be construed as anything but dangerously sinful?

God or mammon?

> No one can serve two masters; for a slave will either hate the one and love the other, or be devoted to the one and despise the other. You cannot serve God and wealth.
> —Matt. 6:24

> If we believe it is morally wrong to wreck the planet, then it follows that it's also morally wrong to profit from wrecking the planet.
> —Author and environmental activist Bill McKibben

Divestment debates force us to examine our use of money, an often painful yet important exercise. In churches, which are often cash-strapped and face many urgent needs, the topic of fossil-fuel divestment can be acutely uncomfortable.

For Christians, however, difficulty has never represented a warrant for avoiding an issue of dire urgency. If faith communities wish to retain their moral legitimacy, they must at the very least debate vigorously the ethics of profiting from an industry which, in the midst of the undeniable past benefits it has generated, is now wreaking literal havoc on God's people and God's creation.

Now!

> To another Jesus said, "Follow me." But he said, "Lord, first let me go and bury my father." But Jesus said to him, "Let the dead bury their own dead; but as for you, go and proclaim the kingdom of God."
> —Luke 9:58-60

Come now, you who say, "Today or tomorrow we will go to such and such a town and spend a year there, doing business and making money.". . . Instead you ought to say, "If the Lord wishes, we will live and do this or that." . . . Anyone, then, who knows the right thing to do and fails to do it, commits sin.
—James 4:13-17

The scientific consensus is clear that climate change represents an urgent problem, and that a rapid response is required. Christians know that, in certain circumstances, Jesus calls us to act with decisiveness. This represents yet another argument in support of divesting and reinvesting. If we believe it is wrong to profit from the fossil-fuel industry, then we should divest—now. If we believe that investing in a clean-energy future is an urgent imperative that will save lives, protect future generations, and increase the well-being of the entire human family, then we should act—now. In circumstances of great consequence, delay is sinful.

For Christians, the concept of *kairos* is relevant here. *Kairos* is a Greek term from the field of archery that describes the tiny window of space and time through which an archer must fire an arrow to hit a moving target. *Kairos* is "an opening, or 'opportunity' or, more precisely, a long tunnel-like aperture through which the archer's arrow has to pass. Successful passage of a *kairos* requires, therefore, that the archer's arrow be fired not only accurately but with enough power for it to penetrate."[12] In Christian theology, *kairos* refers to the supreme timeliness of Jesus' incarnation, death, and resurrection. Like all concepts related to Jesus, *kairos* is not a historical term only but, rather, a living theological reality. In relation to climate change, now is our *kairos*. We must act decisively now. Delay equals death.

Religious Institutions and Divestment

Divestment has a long history in religious communities. Over the past century, Jewish, Christian, Hindu, Muslim, and Buddhist groups have divested from industries including tobacco, alcohol, weapons manufacturing, prison construction and management, gambling, financial-service firms, pharmaceuticals, adult entertainment, and more. Religious groups have also divested from companies operating in countries whose governments systematically abuse human rights—such as apartheid-era South Africa or Sudan in the 2000s.

There are three reasons that religious groups have divested, or seriously debated divesting. The first reason has to do with responding to intentional, grave, and large-scale harm. Religious groups have chosen to divest if a company or industry regularly and intentionally undertakes activities that injure or kill large numbers of people, often people who are vulnerable or marginalized. They've divested because they come to believe that investing in these industries violates the essence of their identity.

For example, when the Union for Reform Judaism[13] (with other religious denominations) voted in 1985 to divest from South Africa, it cited the blight of apartheid and its devastating, systematic impact on South African blacks as its rationale. "In Jewish religious tradition, as well as in Jewish historical experience, racism is an ultimate evil. . . . Apartheid is a system that affronts the most profound values of humanity . . . and violates the teaching of our religion.[14]

Intentional, large-scale, systematic, grave harm confers divestment's moral legitimacy. Despite the benefits that fossil-fuel energy creates, there is no question that fossil fuels are creating such harm.

The second reason has to do with responding to intractable resistance. Investment-ethics experts agree that divestment is a last resort, to be undertaken only after multiple change efforts, including legislative, regulatory, and shareholder advocacy, have failed. These approaches have often helped change many harmful corporate behaviors; all must prove fruitless before divestment should be debated.

Not all industries or corporations are responsive to calls for change, however. And when pressured, some companies or industries seek to downplay the dangers posed by their products by promulgating dishonest research, withholding damaging information, and lobbying against legislation and regulation. The tobacco and fossil-fuel industries have both invested substantially in such efforts, to ill effect. Such a pattern of behavior, exercised over a period of decades, leads some religious groups to consider divesting. An industry's sustained and intractable refusal to change is a second cornerstone of divestment's raison d'etre.[15] Clearly, the fossil-fuel industry meets this criterion.[16]

A final reason religious groups divest is that they view divestment, and the public engagement it involves, as a way to redefine society's moral code. From this perspective, divestment is intended to delegitimize the target industry and to demonstrate its unacceptable repugnance. This is a fundamental part of the process of social change. Very few changes of significance happen without a meaningful level of disgust being attached to the status quo.

This aspect of divestment is largely interpersonal. Within religious groups, divestment involves numerous conversations between group members and leaders, an interactive process of consultation, discussion, and voting that plays out both in public and private arenas. These discussions force leaders to go on record in regard to the issue at hand. If they speak callously, or fail to demonstrate adequate concern, they may lose face and suffer reputational harm. As this process is repeated numerous times in multiple institutions, a new ethos can emerge.

The South African divestment campaign in the 1980s represents an example of this process in action. Through diligent organizing, campaigners succeeded in making apartheid and divestment an unavoidable topic of conversation in homes, on campuses, and within faith communities. Institutions that refused to take this conversation seriously were increasingly seen as morally shallow. Through this process, and in a manner difficult to measure, divestment helped to undermine apartheid's legitimacy and hasten its abolition. The scope of divestment debates, which far exceeded the actual number of divesting institutions, played a role in shifting cultural and political dynamics. South African Archbishop Desmond Tutu says, "In South Africa, we could not have achieved our freedom and just peace without the help of people around the world, who through the use of non-violent means, such as boycotts and divestment, encouraged their governments and other corporate actors to reverse decades-long support for the Apartheid regime."[17]

This aspect of divestment is an imprecise one whose impacts are neither cleanly measurable nor predictable. Almost by definition, these debates create painful disagreement between members of faith communities. But such debates have, at times, created a moral turning point within society by making it impossible for leaders to ignore the brutality of the status quo. In addition, these debates have emboldened or pressured political leaders to take steps to address issues which they'd previously avoided out of fear or indifference.

Divestment and Reinvestment in Action

Since 2012, fossil-fuel divestment resolutions have been promulgated in numerous Christian bodies around the world. Here is a brief summary of some of the most significant examples of action to date.

The Church of Sweden

Before Bill McKibben helped make fossil-fuel divestment a household term, the Church of Sweden had begun to act. In 2009, following a conference convened by the church's archbishop focused on the threat of climate change, the Swedish Church began the divestment process.[18] Notably, it also actively sought out opportunities for investment in the clean-energy sector, providing an admirable example. Its work represents the longstanding commitment of many leaders within the Church of Sweden, particularly Archbishops Anders Wejryd and Antje Jackelén and the Rev. Henrik Grape.

New South Wales Synod of the Uniting Church in Australia

In May 2013, the New South Wales Synod of the Uniting Church in Australia voted that the "Synod should divest from corporations engaged in the extraction of fossil fuels and move instead to investing in renewable energy stocks."[19] The synod's detailed resolution noted the "clear evidence that the threat of climate change is not being adequately addressed by our state and federal governments, or the international community; . . . that rapid expansion of fossil fuel mining (particularly coal and coal seam gas) in Australia is directly threatening agricultural land, human health and biodiversity; . . . [and] that to avoid more than a 20% chance of global temperatures rising beyond the 'extreme danger benchmark' of 2 degrees, 80% of the known coal, oil and gas reserves will need to remain untouched."[20]

United Church of Christ—USA

In July 2013, the U.S.-based United Church of Christ (UCC) passed a resolution at its national General Synod that included a path to divestment from fossil fuels. The resolution targeted United Church Funds (UCF), an investing agency of the denomination which manages the endowments of UCC congregations, conferences, and regional bodies, and the national church's administrative offices. The resolution, a compromise developed through lengthy negotiations between denominational activists and financial managers, required UCF to offer a fossil-fuel-free investing option by the end of 2014, and to conduct regular evaluations of the fossil-fuel holdings in the remainder of its investment holdings, selling off by 2018 those which do not meet the definition of "best in class." The Rev. Jim Antal, the leader of the denomination's pro-divestment activists, has said that he believes this represents an "oxymoron," and that full divestment will follow.[21]

Quakers in Britain

In October 2013, saying that "investing in companies which are engaged in fossil fuel extraction is incompatible with their commitment made in 2011 to become a sustainable low-carbon community," Quakers in Britain adopted a "minute" which committed the denomination to divest from fossil-fuel holdings and to reinvest in renewable energy. "We want to invest in renewable energy and energy-saving schemes. Action we will take as individuals, as meetings and as Britain Yearly Meeting Trustees should aim to minimize damage and strengthen our advocacy position." The minute continued, "We have expressed our difficulties, especially since we all depend in many ways on fossil fuels, but we need to make positive steps towards the change we want to see."[22]

Anglican Church in Aotearoa, New Zealand, and Polynesia

In May 2014, the Anglican Church in Aotearoa, New Zealand, and Polynesia became the first province in the worldwide Anglican Communion to vote to divest from fossil fuels. Noting that "the science of climate change is less politicized in New Zealand than it is elsewhere," New Zealand Anglican divestment leader Dr Matheson Russell noted that the proximity of New Zealand to Pacific island chains at immediate risk from rising sea levels and severe storm activity, combined with the fact that the fossil-fuel industry is not a major part of the New Zealand economy, helped facilitate the decision.[23]

Unitarian Universalist Association—USA

In 2013, in the face of increasing interest in divestment in religious communities, Unitarian Universalist Association (UUA) treasurer Tim Brennan publicly expressed opposition to divestment. An experienced shareholder activist, he noted that divestment would eliminate the denomination's ability to interact with the industry's management. Others expressed concern that divestment would decrease the value and returns and increase the volatility of the UUA's investment holdings.

Over the fall and winter of 2013–14, a group of UUA activists worked with Brennan to craft a compromise resolution that addressed both divestment and reinvestment. The resolution called for divestment from fossil-fuel holdings, with several qualifications. First, the resolution allowed the denomination's to retain its current fossil-fuel holdings if the UUA is already conducting shareholder activism. New fossil-fuel-related holdings may be purchased only for the purpose of shareholder activism—and only at the minimum level required for such advocacy. The resolution also called for investment in companies that would "support a swift transition to a clean energy economy." This resolution was passed in June of 2014 at the UUA General Assembly.[24]

Union Theological Seminary—USA

In June 2014, the trustees of Union Theological Seminary, a leading liberal Protestant seminary in New York City, voted to divest the seminary's USD $108 million endowment from direct holdings of fossil-fuel companies and instructed the managers of their co-mingled funds to transition toward fossil-fuel-free investments.[25] The seminary's president, the Rev. Dr Serene Jones, wrote, "As vulnerable communities have been swallowed by rising shorelines, as potable water has become a commodity of increasing rarity, as hundreds of thousands of people have been killed by violent weather, it is ever clear that humanity's addiction to fossil fuels is death-dealing—or as Christians would say, profoundly sinful." She continued, "We realize that our endowment alone will hardly cause the fossil fuel giants to miss even half a heartbeat. That said, it is on moral grounds that we pursue divestment, and on theological grounds that we trust it matters."[26]

University of Dayton—USA

The University of Dayton, a Catholic, Marianist University, announced in June 2014 that it would divest its USD $670 million endowment from fossil-fuel holdings. In its announcement, the university stated that its divestment would "occur in phases. The University will initially eliminate fossil fuel holdings from its domestic equity accounts . . . then will develop plans to eliminate fossil fuel from international holdings, invest in green and sustainable technologies or holdings, and restrict future investments in private equity or hedge funds whose investments support fossil fuel or significant carbon-producing holdings."[27] "This action . . . is consistent with Catholic social teachings, our Marianist values, and comprehensive campus wide sustainability initiatives and commitments under the American College and University Presidents' Climate Commitment," University of Dayton President Daniel J. Curran said.

". . . Our Marianist values of leadership and service to humanity call upon us to act on these principles and serve as a catalyst for civil discussion and positive change that benefits our planet."[28]

A listing of all known faith-based fossil-fuel divestment efforts can be found on GreenFaith's website.[29] The list grows consistently, and includes Christian communities in Europe, Canada, the USA, Australia, and New Zealand.

World Council of Churches, Geneva, Switzerland

In July 2014, the World Council of Churches (WCC) endorsed fossil-fuel divestment and agreed to phase out its own holdings of fossil-fuel investments and encouraged its members to do the same. The WCC is a fellowship of over 300 churches which represent 590 million people in 150 countries, and its announcement has been viewed by divestment advocates as a substantial affirmation of wider divestment efforts, though the WCC's decision applies only to its own investments, not those of its members. The World Council of Churches Financial Policy Committee reported that it "discussed the ethical investment criteria, and considered that the list of sectors in which the WCC does not invest should be extended to include fossil fuels."[30]

Dr Guillermo Kerber, who coordinates the WCC's work on care for creation and climate justice, noted that the WCC's "general ethical guidelines for investment already included the concern for a sustainable environment, for future generations and CO_2 footprint. Adding fossil fuels to the list of sectors where the WCC does not invest in," he continued, "serves to strengthen the governing body's commitment on climate change."[31] Kerber also noted that youth leaders initiated the divestment conversation within the WCC.

Diocese of Exeter

In late 2014, the Diocese of Exeter became the first diocese in the Church of England to vote to divest, and called on the Church of England itself to divest. The Exeter resolution called for disinvestment from coal and tar sands "at the earliest opportunity," from oil in three years, and from natural gas in five years.[32]

Numerous religious groups continue to join the divestment movement, which shows no sign of slowing. Indeed, the momentum behind the divestment movement within the faith community has grown quickly and impressively, and is likely to continue to do so.

Reinvest

To maintain integrity in the face of climate change, however, religious groups cannot just divest from fossil-fuel holdings. At a minimum, they must make their own facilities more energy efficient and make widespread use of renewable energy. This isn't the 1960s, or even the '80s. Protest and divestment alone aren't enough. To maintain integrity, religious groups need to reinvest, to embody a sustainable future.

For religious groups, modest investments in making their own facilities energy efficient pay rapid returns, initially far greater than those earned in the stock market. There are a growing number of ways religious groups can partner with energy-service companies and renewable-energy developers to adopt conservation and efficiency measures and renewable energy at their facilities, and to encourage their members to do the same.

To date, religious groups have, at best, a modest track record in this regard. This needs to change. There is hope in the fact that reducing religious institutions' carbon footprints is very achievable, and also creates financial and mission-related benefits. A 2012 report from the Sustainable Endowments Coalition shows that colleges that invested in energy-efficient lighting and heating in their own facilities saw a median return on investment of 28 percent.[33] GreenFaith's experience conducting energy efficiency and conservation work with hundreds of religious sites has taught us that similar returns are easily within the reach of most religious institutions.

Religious groups should identify opportunities to invest in the development of a clean energy future—particularly in the developing world. A growing number of experts can identify clean-energy projects suitable for religious investors. With research, religious groups could jointly identify a pool of such investments. A multifaith team of denominational representatives could craft such a strategy, use the positive attention it would generate to burnish their own reputations, exert more credible pressure on political and corporate leaders, while also saving lives.

Just as some have questioned the concept of divestment, others will raise concerns about reinvestment. They will argue that religious groups should not try to "pick winners" in the fast-changing clean-energy sector, or that upgrading religious facilities is a capital expense, not an investment. We disagree. Opportunities can be created for religious groups to invest in pools of clean-energy businesses. And regarding the question of capital expense versus investment, if religious groups have such a financially rewarding, ethical use for their funds, they should do so in a heartbeat.

Conclusion

There comes a time when a pattern of recklessly dangerous behaviour reaches a critical point. The climate crisis represents such a point. The data could not be clearer about the titanic magnitude of the danger posed by climate change. To date, society's refusal to address this issue reflects an avoidance of a fundamental human responsibility. Our current path represents a cataclysmic mistake.

We have reached a time when new action is required. Action beyond legislative advocacy and shareholder advocacy. Action that reflects the gravity of the situation and the legitimate desperation of our "dark interval." Action based on the belief that integrity requires us to act firmly, sacrificing a measure of our own security on behalf of the future of our children and the planet. Action that firmly rejects the status quo and equally firmly embraces a future for life.

Fossil-fuel divestment and reinvestment in a clean-energy future are a moral, religious responsibility. Religious groups must engage this debate with a fierce loyalty to God and God's compassionate, just love for humanity and all creation. For those who agree—it is time to act.

And for those who disagree but who affirm that climate change poses a severe threat, we have one question. If not divestment and reinvestment, then what would you propose to force the necessary response to the climate crisis?

CHAPTER 17

Indigenous Ecological Spirituality and the Land

Tourism and Development in Northeast India

Marina Ngursangzeli Behera

The northeastern region of India[1] **is considered the green belt of India,** being rich in natural resources—oil, natural gas, minerals, and, most importantly, valuable forests. Northeast India (NEI) was once considered an environmentally friendly region with its copious rainfall every year, heavily forested mountains and jungles, and clear running rivers. Today the region is continuously and ruthlessly subjected to the degradation of its environment due to the consequences of the tourism industry.

The NEI region has been ignored both politically and economically for a long time. Today, there is serious interest being shown in the region. A Ministry of Development of North Eastern Region (DoNER) was established in 2001, specifically to oversee socioeconomic development. The tourism potential of the NEI region is now recognized and acknowledged by the Indian government and the neighbouring countries that border this region. The tourism industry, which has become a major component of economic globalization and is considered the largest industry in the service sector, continues to seek out the last "Shangri-las" in the once-most-isolated places, such as Northeast India.

With "nature" and "culture" being today's catchwords, more travellers have set out in search of exotic cultural experiences that southern countries have to offer. India intends to be a major player in this market by upgrading the existing tourist spots and developing new ones. The NEI region, which had been politically ignored in the past, has suddenly become very prominent in the central government's plan.

The overall developmental programs that have been proposed and initiated by the central (federal) government and its partners, like the Asian Development Bank (ADB) through the South Asia Sub-regional Economic Cooperation (SASEC) for the NEI region, identifies tourism development as the priority. This is clearly reflected in the Vision 2020 for India, where the development of core competencies in areas in which the region has apparent comparative advantage was alluded to as a means to deal with the impact of globalisation, namely economic or trade or market forces.[2]

Tourism has several adverse effects on climate change and the natural environment. This holds true for the NEI region as well as other places where the indigenous people's culture and their habitat—their land—are being commodified as goods to be sold in the globalized tourism market. Himanshi Dhawan, in her *Times of India* article "Northeast emerges as the new hotspot for tourists," writes, "The number of tourists, both domestic and international, heading for the northeast has gone up dramatically in the past year. . . . Till recently, these states, beautiful and unexplored as they were, rated low on tourist options. . . . But effective initiatives by local governments, travel companies and airlines have changed all that."[3]

This essay deals with the phenomenon of tourism development in the NEI region and its impact on the indigenous people, their land, and the environment in the region. It must be noted here that tourism development thus far has been viewed and encouraged from a governmental and economic point of view and the church in NEI, along with local communities, have spent little time studying this or its implications. There have been no debates or discussions held in public forums to ascertain whether changes that tourism will inevitably bring will benefit local communities.

Programs/Projects for the Development of Northeast India

Before the recent tourism promotion, the process of economic development had been slow in the NEI region for many reasons. Traditional systems of self-governance and social customs of livelihood had remained virtually untouched during the British rule. The creation of the rail network for commercial purposes, linking the areas where tea is grown, was the only major economic activity taken up during the period.[4] Special attention has been given to the economic development of the region from the Eighth Five-Year Plan onward. In October 1996 the central government announced "New Initiatives for the North Eastern Region" and proposed a number of measures to develop NEI. Emphasis was given to connectivity as the key area for development of NEI and stated how transport infrastructure would have a special thrust. Interestingly, tourism was positioned as one of the critical parameters of this regions' growth. The document also pointed out how the Look East Policy (LEP)[5] is important for breaking these areas out of isolation and to gradually develop the region into an economic identity and move it on a higher growth path.[6]

The LEP and closer interactions with ASEAN (Association of South East Asian Nations) have accrued huge economic benefits for India. The India-ASEAN trade has grown by 23 percent in the last ten years and, in 2012–13, the total trade between India and ASEAN was worth USD $80 billion. It is expected that with the signing of the Agreement on Free Trade Area, the total bilateral trade will increase to USD $100 billion by 2015 and USD $200 billion by 2020. The NEI region, which is the gateway to the Southeast Asian countries, has not really benefited from the opportunities opened up through the LEP.[7] Development of communications and transportation such as road, rail, and air links in the Northeast are considered a strategic means to overcome this problem and also as a means to help the NEI region integrate with the rest of the country. High-visibility projects that could be implemented in a short period in the fields of health, culture, tourism, capacity building, and food processing have also been considered desirable. Countries such as Singapore, South Korea, Japan, and Taiwan would be welcomed in the Northeast for capacity building in areas such as agriculture, tourism, hotel development, and bamboo technology.[8]

Tourism Development in the Northeast India Region

The NEI region is a major focus under various schemes of the Central Ministry of Tourism. It encourages and funds the states in this region for the development of infrastructure in the identified destinations and circuits.[9] Tourism has become the new development slogan for the northeast states of India and an important component of the government's ambitious plan for development of the country's northeast region. The country's Ninth Five-Year Plan (1997–2002) specifically focussed on exploiting the tourism potential of the northeast region by promoting eco-tourism and adventure tourism and improving infrastructure, accommodations, and tourist-guide facilities. The National Tourism Policy of 2012 also recognizes the need for promoting specific forms of tourism like river-cruise tourism, eco-tourism, and village tourism in NEI.[10]

The Vision 2020 document, released by Prime Minister Manmohan Singh[11] in 2008 for the northeastern region, clearly focuses on developing services that will help to accelerate development with a clear recognition of the considerable potential for expanding tourism.[12] It even goes as far as to say that "tourism [is] a prime and over-arching activity for NER States."[13] The long-term vision, apart from developing connectivity and communication, is for both the central and state governments to prioritize and evolve a blueprint for a common market and a common administrative and economic unit for the region.[14] There is an emphasis on the importance of looking at the region as a whole, offering tourism that cuts across state boundaries and the need for the region to "rebrand itself" along these lines; indeed, it goes as far as to say that the slogan "Paradise Unexplored" with regard to the NEI inspires "instant recall" in the mind of an alert traveller.[15] In 2008, the NEC (North Eastern Council) formed the North Eastern States Tourism Forum (NEST) with representation from the various states to coordinate intrastate and interstate tourism plans, such as the consolidation of circuits and marketing.

Emphasis has been placed on having all the NEI states improve their tourism-related infrastructure and to relax the permit system in the region, and different state governments have taken several initiatives in their respective states. The Assam Development Report has recommended providing fiscal incentives to private entrepreneurs for the marketing of potential tourist places. The Sikkim State Development Report recommends opening up the Nathu La pass to boost tourism in the region.[16]

NEI has also been promoted as a destination to "Leave Travel Concession" (LTC) travellers, which is not much welcomed by the indigenous people. The protest and appeal letter against the proposal that was sent to the minister of DoNER highlights the many adverse implications for the indigenous people of the region and their environment.[17]

The Impact of Tourism Development on the Indigenous People, Their Land, and the Eco-system

One of the immediate adverse impacts would be the subjection of the NEI region to intense development, not just in terms of tourism projects, but infrastructure development such as roads and airports. Almost all of the identified areas are ecologically and socially sensitive regions whose capacity cannot sustain unlimited use of space and resources for tourism and infrastructure. These would lead to natural resources being exploited, communities displaced, the destination spent and overexposed, so that the pristine and unspoiled attraction of the region, as has been aggressively marketed, would soon lose its meaning.[18] The development of roads to tourist destinations results in the felling of trees and loss of forest and degradation of the land. Developing and increasing airline services to cater to the tourists travelling to the region adds to the world's carbon emissions and contributes to global warming. The massive use of fossil-fuel-based energy, which would be required for the lighting and air conditioning of hotels, resorts, and guest houses, also needs to be taken into account.

The proposed activities such as trekking, river-based tourism, and adventure tourism are not at all suitable for the NEI region and would have grave adverse impacts on the ecology. For instance, to use the Ganga-Brahmaputra basin for river-based activities like cruising would result in the pollution of the river in the form of sewage effluents and oil and fuel leakages from the cruise ships, and many other forms of organic and inorganic waste. There is also the potential harm for the various lifeforms in the river. Adventure tourism like rafting and mountaineering can also heighten the tension on the eco-system.

Under the myth of sustainable tourism[19] and eco-tourism,[20] the environment, natural resources, the eco-system, the regional ecology serving as the backdrop to commercialized tourism are being continuously exploited. Branded by the tourism industry as "Paradise Unexplored," NEI is slowly turning into a place where the biodiversity of the region is threatened, where pollution is on the increase, where agricultural land and sacred spaces are being appropriated for the tourism industry, and where natural resources such as potable water and energy resources are diminishing. The exploitative, developmental model of tourism does not serve the indigenous people of the NEI at all, but only leads to the exploitation of them and their land. Tourism development, even though claiming to be eco-tourism, fails to develop responsible measures even in terms of proper waste and plastic management, which is a serious affront to the indigenous communities in the NEI region who feel pride in the care of their environment. A question asked by a participant at a consultation held in Shillong on "Churches' Response to Development and Tourism" in 2011, "When the mountain disappears, what will be our identity?,"[21] is something that is worth pondering.

A report in the *Times of India* on 27 September 2013, titled "World Tourism Day observed in Nagaland,"[22] clearly positions the model of development for the NEI region that the central government had drawn up as directly relating to tourism development in the region. It is ironic that even the projects to meet the need for pure drinking water in the region are connected with the needs and demands of the tourism market. It is logical to wonder whether all the development the central government is proposing for the NEI region is being done because of genuine concerns for the people of the region.

One needs to take into account that it is not only the land, nature, and the biodiversity of the NEI region that are being threatened and suffering under the tourism development, but the indigenous people themselves. Not only have the indigenous people along with their culture—their very way of life—been imaged, packaged, and sold as commodities in the global tourism market, but this has also led to the destruction of ancient and sacred cultural ties of the people to their land. They are gradually being alienated from their land, which for them is the basis of their subsistence, identity, and spirituality.

Indigenous traditional land systems are often ignored, and biodiversity is lost in the process. The promotion of the sacred groves in the region as tourist attractions is a good example. In an age-old practice, local communities protect certain areas of land near the villages through customary taboos and sanctions with cultural and ecological implications. They are a part of the cultural heritage of the indigenous people. Sacred groves also contain several medicinal plants used in *ayurveda*, tribal, and folk medicine.[23]

Among the many types of sacred groves, there are those in Meghalaya which have been considered traditionally sacred and therefore, ideally, not even a fallen twig could be removed from within the grove, where a whole variety of religious beliefs and practices are grounded. Apart from the sacred groves functioning as the medium through which the indigenous people express their relationship with the divine and nature, they also have an important socio-cultural and sociopolitical function. They provide cultural space to the community as a common property and also for the assertion of group identity and solidarity. Displacement from or violation of these lands can amount to cultural devastation.

Tribal villages have become showcases for visiting tourists. Consider, for example, the Khonoma Village in Nagaland, a village of 500 houses, furnished with a grant from the Ministry of Tourism and Culture in New Delhi and supported by the Nagaland state government, which is bent on turning the village into a showcase eco-tourism model. The village is awash with plans and dreams that revolve around the mantra "When the tourists come," repeated by the younger members of the community. This illustrates the pragmatic approach of the village youth toward the potential that tourism may hold for them. It is a perspective shared by the village council, which chooses to see tourist revenues as one among several streams of communal income, and certainly not the most important, despite the beliefs of distant planners in New Delhi and opportunist bureaucrats in Kohima. [24]

The promise of the local people benefiting from the tourism industry is but a myth. The promise of employment and jobs and the creation of a new identity only results in their being identified and constructed as those who should serve and cater to the needs and pleasures of the tourists, their masters, whom they are expected to please.[25]

Tourism development not only leads to the physical alteration of the landscape and the environment through human activity that degrades the quality of the water or the air, alters or destroys vegetation, increases noise, and affects wildlife. Experiential degradation can also occur as a result of increased human recreational activity, even when no substantial physical alterations occur. There also occurs simultaneously a perceptional alteration. For example, the sacred sites and religious sites of the indigenous people that form an intrinsic part of their identity and culture are now being sold as recreational centres for tourists. This change may be as devastating as actual physical degradation for the indigenous people in the NEI region.[26]

Conclusion

For the indigenous people of NEI who belong traditionally to agrarian societies, land is central to their life and existence as a community. Land is the source of life and is also sacred. The land provides them with space, food, and many of the basic resources to meet their everyday needs. The land and their identity are intertwined in terms of their identity as a community, their social system, culture, values, and their economic well-being. Traditionally, land has never been considered as a private or individual property; it belongs to a higher supreme being or power who created the whole cosmos, under whose care and protection human beings are allowed to partake of the resources or the produce of the land, and only in conjunction with the understanding that it is a taboo to abuse the land and the natural resources by going beyond a limit that meets their daily needs. The people have only understood themselves as stewards and caretakers of the land and its natural resources. It was through the land that they interpret what life means, and learn about the divine and the ecological order and their interconnectedness with the rest of creation.

Today, tourism, promoted by the neoliberal globalized market economy, has ruthlessly and aggressively appropriated the land of the indigenous people, constructing it and imaging it as something to be sold as experiences of pleasure in the globalized market. It has torn away and alienated the people from their land, resulting in the loss of identity and the loss of space where they had communion with the divine and the whole cosmos. There has been diversion of people's essential resources such as land, water, electricity, and other infrastructure to entertain tourists and support hotels, resorts, golf courses, and amusement parks. In opposition to the exploitative model of development that has been implemented through the tourism industry, there is the need to recapture the indigenous cosmology and develop an alternative model or framework that is more cosmic-centred and rooted in the integration and interrelatedness of all human beings with the land and the rest of God's creations.

There is the need to rediscover the purpose of our lives within the context of God's larger purposes for the world and to exercise our stewardship in a more embracing and holistic manner. We are to be stewards not only of resources that generate sustainability but also in ways that sustain life for the poor and vulnerable. In biblical terms, we are to act out God's compassion to those who are being oppressed by the unfair market economy.

It is important to look beyond the propaganda for national integration and the model of development that the Indian central government, with its partners, has proposed for the NEI region. One needs to understand the implications all these plans have for the indigenous people of the region and the environment and whether they really fit in with what is being promised in the name of development. With the economic liberalization and India's LEP since the early 1990s, one aspect of development saw foreign investment and international financial institutions being granted increased access to the region. However, this had adverse implications for the unique structures, traditions, and resources and the ecology in the NEI region. Hence, it is important to deal with this sort of development with caution.

The church in the NEI region needs to open its eyes and take seriously the emphasis on the development of tourism, with its implications for the people and their land and for the ecology of the region. It cannot remain aloof but must play a leading and constructive role in looking after the welfare of the people as well as remain true to its prophetic calling. It is a role that the church must consider urgently. It needs to critically study why there is a sudden interest in opening up the NEI region in the name of development, which previously was a restricted area. The church needs to seriously consider and ask who will benefit, and at what and whose cost, with the opening up of the NEI region. The church needs to take steps to educate its members and prepare the people for the inevitable impact of the development of tourism. The church will need to find resources within it to address the implications of tourism in the region, while staying rooted in its faith in a God who is for justice, peace, and the integrity of all creation.

CHAPTER 18

The Dance of Krishna

Environmental Challenges in India and Hindu Resources

Siddhartha

> Krishna brought forward the cows and played on his flute through the forest of Vrindavan, which was full of flowers, vegetables and pasturing grass. The Vrindavan forest was as sanctified as the clear mind of a devotee and was full of bees, flowers and fruits. There were chirping birds and clear water lakes with waters that could relieve one of all fatigues. Sweet flavored breezes blew always, refreshing the mind and body.
>
> —Srimad Bhagavatam[1]

When we look at the way nature is being polluted in today's India, the above passage on Krishna in Vrindavan might appear utopian, but we really do not have to go back into the mists of early history to bear out this statement. Even as little as 50 years ago, much of India was environmentally healthy, for the exploitation of nature at the altar of consumer society was yet to take off in a concerted way. Poverty was, of course, endemic then, and hunger and starvation were widespread.

During my college days in Tiruchirapalli, I vividly remember an old woman picking up rice from a drain outside our dining hall. The rice came from the fingers of the students when they washed their hand after a meal. The water flowed down a drain and collected at the very end, where the old woman devotedly picked the rice grains, one by one. I also remember a bus journey I made in the state of Odisha with a sling bag at my side, which contained a couple of bananas. I noticed that the bananas were totally crushed and when the bus made a stop at a village I flung them through the window into a paddy field. In a moment there was a flutter and a few women on the roadside rushed into the field to retrieve the soggy bananas from the inundated paddies. Such sights may be rare now, in an era when food grains are distributed by the government to the poor at a minimal price.

In my college days I was part of a left-wing group that believed that we had to work toward structural change in society to arrive at a measure of social justice. The world was simpler then, and we believed that it was possible to see the elimination of poverty and injustice in our lifetime. Today we find that injustice still continues and the gap between the poor and the rich has widened dramatically not only in India, but all over the world. In addition to the poor and minorities, we find that women continue to be marginalized and oppressed.

In light of all these problems, nothing prepared us for the advent of the climate crisis. In a study completed in 2013, Professor Steven Sherwood from the University of New South Wales in Australia says that the planet will see an increase in temperature of 4 degrees Celsius by the year 2100 if greenhouse emissions are not reduced.[2] This is twice the level climate scientists regard as dangerous.

Even the notion of environment was not part of our vocabulary in my student days. Development was seen as positive, provided there was equity. We now find ourselves in a somewhat confused period where some are even stating that growth-oriented development is at the root of our climate woes. A few have proposed that we get rid of the notion of development altogether and look at sustainable futures within a paradigm of zero growth. It is not always clear what this really means and how to bring about the necessary changes without rocking the boat perilously. Many argue that such alternatives are utopian and that the best we can hope for is to reform within the existing system, although that is clearly not enough to prevent the impending catastrophe.

Climate-Change Implications for India

Coastal eco-system

> That which, resting in the waters, is other than the waters; which the waters do not know; of which the waters are the body; which controls the waters from within: this is your self, the inner controller, the immortal. (Brhadaranyaka Upanishad III.7.3–23)

Unfortunately, sufficient studies to project the impact of climate change on fish populations in India's coastal waters have not been done so far. But figures show that the annual per capita fish catch by artisanal fishers in the country dropped from 2,590 kilograms in 1980 to 420 kilograms in 1996–97. This is due to a combination of climate factors, chemical pollution of the sea-waters, and fishing with mechanized boats and trawlers. The livelihood of small fisherfolk has been badly affected as fish stocks dwindle. For example, the state of Maharashtra's annual fish catch has declined from 4.5 lakh tons in 2002 to 2.25 lakh tons in 2010,[3] and artisanal fish workers are the most seriously affected by this.

Sardines and mackerel are a cheap source of protein and provide nutritional food for millions of coastal people. E. Vivekanandan states that these small fish, especially the oil sardine, were known for their restricted distribution between latitude 8°N and 14°N and longitude 75°E and 77°E (Malabar zone along the southwest coast of India), where the annual average sea surface temperature ranges from 27 to 29 degrees Celsius.[4]

But, as temperatures go up, fish are migrating from warmer to cooler waters. They move to cooler latitudes or, alternatively, descend deeper in the sea. Sardine and mackerel are found to be migrating from the Malabar coast to cooler temperatures further north, beyond latitude 14°N. The Indian mackerel lives close to the sea surface and has been traditionally caught by artisanal fishers using surface-drift gillnets. Rising temperatures mean that the fish is now largely being caught by large bottom-trawling boats. Fish are known to mature faster with higher temperatures. They also lay fewer eggs, bringing down production; fish larvae are also vulnerable to temperature changes.

Fish populations in the coastal waters are facing the usual problems of over-fishing, pollution (garbage, including plastic waste, and chemical waste), and habitat degradation. Catching of juvenile fish also reduces the fish population substantially. Fish stocks are also affected by changes in species composition of phytoplankton at higher temperatures.[5] Marine phytoplanktons are tiny plants that represent food for fish and suck up carbon dioxide and release oxygen. They are declining rapidly as a result of a loss of nutrients with the warming of seas. Worldwide phytoplankton levels have gone down by 40 percent since the 1940s. Apart from this, the oceans also absorb about 26 percent of carbon dioxide emerging from a variety of human activities. This process happens with the help of different living marine organisms and habitats that include mangroves, salt marshes, sea grasses, and seaweed, frequently referred to as "blue forests."

Rising acidity levels in the seas as a result of climate change have harmful effects on coral reefs and calcium-bearing organisms.[6] Coral reefs support about one million species. In the Indian seas, coral reefs are found in the Gulf of Mannar, Gulf of Kutch, Palk Bay, Andaman Sea, and Lakshadweep Sea. Rising carbon-dioxide concentration is leading to acidification of the sea, damaging these coral reefs. Coral bleaching is already happening in a noticeable way in the Andaman Sea. It is believed that coral reefs will disappear in about 50 years time. About 25 percent of the world's marine species are today found in coral reefs.

In the context of climate change, helping to reduce fish mortality in the majority of overexploited fisheries becomes important. The Indian government should prevent overfishing to ensure that fishing efforts are commensurate with the productive capacity of the oceans. Seasonal closure of mechanized fishing for 45 to 60 days is perhaps the only regulatory measure that is being followed at present. The rights of people engaged in artisanal small-scale fisheries must be protected and preferential access to traditional fishing grounds must be respected.[7] About 1.2 percent of global oil consumption is used in the fisheries chain.[8] Hence, reducing overfishing may help to conserve fish stocks in the context of climate change. The Code of Conduct for Responsible Fisheries and Integrated Ecosystem-based Fisheries Management must be strictly enforced.[9]

Saline water entering coastal farms is negatively affecting the agricultural production of small farmers in coastal areas in India. The situation has become untenable in places like the Sunderban islands. Salt water is also intruding into the drinking water of villagers along the coast. Reports from Kerala indicate that about 200 fishing villages are slowly being battered by the sea. According to the Kerala Swatantra Matsya Thozhilali Federation (Kerala Independent

Fish Workers Federation), about 10,000 people will be rendered homeless along the Kerala- Kanyakumari coast.[10]

As far as beach erosion is concerned, India is ranked as the most vulnerable of the 51 countries that IPCC scientists have studied.[11] The experience of the Maharashtra coast shows that warmer seas might lead to more cyclones and storm surges.[12] Bangladesh and India account for 86 percent of the mortality from tropical cyclones and severe storms.[13]

Forest communities and climate change

Krishna had this to say of Vrindavan forest:

> Just look at those most fortunate trees of Vrindavan. They have dedicated their lives to the welfare of others. Individually they are tolerating all kinds of natural disturbances, such as hurricanes, torrents of rain, scorching heat and piercing cold, but they are very careful to relieve our fatigues and give us shelter. My dear friends, I think they are glorified in this birth as trees. They are so careful to provide shelter to others that they are like noble, highly elevated charitable men who never deny charity to one who approaches them. No one is denied shelter by these trees. They supply various kinds of facilities to human society, such as leaves, flowers, fruit, shade, roots, bark, flavor extracts and fuel. (Srimad Bhagavatam 10.22.32–35)

Forest and tree cover in India accounts for 78.37 million hectares (Mha). This represents 23.84 percent of the geographic area of the country. The forest cover alone is 69.09 Mha or 21.02 percent of the geographic area. India has only 2 percent of the world's forest cover, but includes 16 percent of the global population and 18 percent of the world's cattle.[14] About 200 million people find their livelihoods from the forests of India. About 40 percent of the country's forest needs and 30 percent of its energy needs come from the forests.[15]

As a result of the pressures of development and global warming, forest ecosystems in India will undergo changes in species composition, productivity, and biodiversity. Like other parts of the world, forest communities and forests in India face major challenges from mining, timber, and land mafia as well as the construction of factories, roads, and railway lines.

In 1974 the well-known Chipko movement had peasant women from Himalayan villages in Uttarakhand hugging trees to prevent them being felled by the contractors. One of the songs sung in the movement narrated a confrontation between a forest officer and the protesting women:

> The forester asks:
> What does the forest bear?
> Resin, timber and foreign exchange.
> The village women reply:
> What does the forest bear?
> Soil, water and pure air.
> Soil, water and pure air
> Are the basis of our life.[16]

The *adivasi* (tribal) populations of the state of Odisha have been struggling to prevent their lands being taken over for mining and industrial purposes. Two giant projects of multinational corporations have been hanging fire as a result of the *adivasi* struggles. The South Korean company POSCO faced long delays in setting up their massive steel plant on *adivasi* lands. Unfortunately, the government of India has given the "go" signal for this plant to take off. As for the Vedanta bauxite mine at Niyamgiri, the *adivasi* struggles appear to have produced results. In January 2014 the central government rejected Vedanta's proposal to clear the forests to start bauxite mining. The state government of Odisha has now appealed to the Supreme Court, challenging this decision.

In many parts of India, forest communities depend on rain-fed agriculture. They also collect non-timber forest products (NTFP), also known as *minor forest produce* (MFP), that include honey, medicinal and aromatic plants (MAP), oil seeds, fiber and floss, resins, edible plants, bamboo, reeds, and grasses. Unfortunately, the forest communities continue to face difficulties due to the government's tardiness in implementing the Forest Rights Act, which gives them ownership rights of land they have traditionally possessed and also the rights to livelihood. Climate change is definitely staring them in the face, as it reduces biodiversity and threatens their food security.

Apart from modern know-how related to forest management, the knowledge of indigenous communities needs to be taken seriously. This knowledge will help these forest communities to adapt to climate change and mitigate its impacts. For example, in Kerala, *adivasi* communities know what varieties of paddy are to be sown when there is too much rainfall and what varieties when there is too little. The recognition of indigenous peoples' rights over their forests and the significance of their role in forest management are of crucial importance in dealing with climate-change impacts.

Forests play a significant role in the earth's carbon cycle. It is well known that trees convert atmospheric carbon from CO_2 into organic woody biomass as part of the process called photosynthesis. The carbon is then stored in trees until the woody biomass is destroyed. Sequestration is the name given to the storage of carbon. Carbon absorption by trees stops when forests are cut down. When wood is burned or decays, the carbon absorbed by trees is released as carbon dioxide. While croplands also store carbon, forests store between 20 and 100 times more carbon per unit area.[17]

In the context of climate change, the United Nations is attempting to help developing countries protect, better manage, and save forest resources through a program called REDD (Reducing Emissions from Deforestation and Forest Degradation). REDD Plus is not only concerned with checking deforestation and forest degradation but also includes positive efforts at conservation, sustainable management, and enhancement of forest carbon stocks. The Forest Rights Act in India is meant to empower forest communities to participate in the use and management of forest resources. However, civil-society groups are wary of the government and suspect that the government will use REDD Plus to weaken the Forest Rights Act.[18]

The urban context

India's urban population is likely to grow to 875 million people by 2050, representing 54 percent of the total population.[19] This is more than the current combined population of the USA, Brazil, Russia, Japan, and Germany. In Mumbai, India's second largest city, about 9 million people, or 62 percent of the city's population, live in slums.[20] By 2017, India's slum population is slated to rise to 104 million people.[21]

In many of the slums there is little sanitation or electricity and inadequate water supply. Often slums are located in low-lying areas that are prone to flooding. I remember the time I had a conversation with poor women in a Bangalore slum when it began to rain heavily. I remarked that the rains would bring welcome relief, as it had been quite hot the past weeks. One elderly woman retorted that for her, the rain meant she would not sleep that night as her hut would be flooded. In coastal cities there is evidence of increased flooding in the past few years. Calcutta city has been ranked as the seventh most perilous urban conglomeration as far as storms, storm surges, tsunamis, and river floods are concerned.[22] In 2005 Mumbai was struck by a cyclone that brought 94 centimeters of rain in 24 hours. More than a thousand people died.[23] The government of Maharashtra developed India's first urban disaster-management plan for the city of Mumbai but very little follow-up action was taken to implement it. Poor housing conditions, no adherence to building regulations and lack of land tenure increase the woes of people in the slums.

The city of Bangalore was known as the garden city of India not too long ago. Today many of the trees have been cut as part of the road-widening program and old houses were knocked down to build apartments. The massive increase in the number of cars, motorcycles, buses and trucks in the city has dramatically increased emissions. Air pollution in some parts of the city is three to four times higher than the permissible limits![24] Crossing the road has become a nightmare for most pedestrians, especially for the older persons. The temperature in the city is about 2 degrees Celsius higher than the surrounding areas, creating heat islands. Of course, in cities like Delhi, the heat islands are more serious with poorer people suffering from dehydration or dying of heat strokes.

Many cities are now getting water from faraway places, often using a lot of electricity to pump the water. Some cities have made rainwater harvesting mandatory. Cities need to be greened, terrace gardens need to be promoted (of course the poor in the slums do not have terraces!), and rooftops painted white to reflect light back.

Municipal authorities have to come up with plans to deal with emissions that come from power usage, vehicle use, and waste management (waste releases huge amounts of methane and other gases). Waste is also taken from the cities and dumped in nearby villages, creating health hazards to the local people.

Climate change and mountainous regions

> This Govardhana Hill is the best! O my friends, this hill supplies Krishna and Balarama, along with their calves, cows and cowherd friends, with all kinds of necessities—water for drinking, very soft grass, caves, fruits, flowers and vegetables. In this way the hill offers respects to the Lord (Shrimad Bhagavatam 10.21.18).[25]

Krishna himself is supposed to have merged with the sacred hill.

Mountain systems represent about 20 percent of the earth's surface. According to an IPCC report, "mountains provide the direct life-support base for about a tenth of humankind and indirectly affect the lives of more than half."[26] In India, mountainous regions provide water, electricity, biological diversity, minerals, forest, vegetables, and other agricultural produce.

There is evidence that temperatures in the Himalayas are rising faster than the global average. Surface air temperatures in the Himalayas have gone up by one degree in the last decade.[27] The Hindu Kush–Karakoram–Himalayan glaciers, which run 2,000 kilometers from east to west and contain more than 60,000 square kilometers of ice, are a major source of water for hundreds of millions of people who live in South Asia. These glaciers provide water to rivers, for agriculture and industry, domestic consumption, and energy production. But, due to climate change, the glaciers are shrinking. The Himalayan glaciers lost about 174 gigatons of water between 2003 and 2009, causing ruinous floods in the Indus, Ganges, and Brahmaputra rivers. Pollution is also contributing to the melting of glaciers, due to an "Asian brown cloud" comprising 2 million particles of soot and dark particles that enter the atmosphere every year due to pollution in India and China.[28]

Due to overexploitation, many mountainous regions in India are losing their forest cover, leading to frequent landslides. In addition, there is loss of biodiversity and loss of medicinal plants. In the past few decades tourism has become a major economic activity, adding to the environment woes. In some mountain areas the increase in the number of people has led to overgrazing of livestock and deforestation.

In June 2013, flash floods in Uttarakhand led to the death of about 1,000 people. Roads, bridges, and other infrastructure were washed away and thousands of homes destroyed. If the region had not been so badly deforested the impact of the floods would have been less pronounced.[29]

Through the network created by Pipal Tree, I have been involved with promoting millet production in the mountainous district of Nilgiris in Tamilnadu state, South India. Through the work of Keystone, a Kothagiri-based organization, we discovered that millet production in the district had declined to a paltry 43 acres. This is largely because of the advent of monoculture in the district, with tea plantations replacing millets. In recent years tourism has grown by leaps and bounds and a large number of hotels and resorts are being constructed. The tea plantations and vegetable gardens are slowly being converted into housing sites.

Climate change is leading to unpredictable and fluctuating rainfall in towns like Udagamandalam, Coonoor, and Kothagiri in the Nilgiris. Even drinking water is becoming a problem in the summer months. Vegetable cultivation is suffering due to frequent drought conditions, impoverishing the small cultivators.

Health and the climate crisis

The increase of droughts, diseases, floods, famine, and of extreme heat and humidity are partly the result of warming temperatures and variable precipitation. The direct and indirect results of rising temperatures are malnutrition, decrease in water availability, spread of diseases like malaria, chikungunya, dengue, and diarrheal illnesses. Dehydration, heat exhaustion, and heat stroke are other health problems which will affect a large number of people, mainly the poorer Indians. The elderly, young children, and newborns are particularly at risk when temperatures rise. People's productivity also goes down with higher humidity and temperatures.

As climate change progresses, floods and cyclones can create loss of life and lead to large-scale health problems. Water and vector-borne and respiratory diseases can become all too common. Eye diseases like cataracts, dry eyes, pterygium, and vernal kerato conjunctivitis will become more common as temperatures rise. Each year people die in India from heat-related problems. Deaths from heat stroke in the year 1998 were 1,658. Most of the cases came from Andhra Pradesh, Odisha, Punjab, Uttar Pradesh, Rajasthan, Bihar, and Madhya Pradesh.[30] In 2003 a heat wave killed 1,421 people in Andhra Pradesh. Extreme weather can have different types of impact on different populations, depending on their ability to cope. Cyclones in Bangladesh caused 300,000 deaths in 1970, whereas when Hurricane Andrew struck the United States in 1992 the death toll was 55.[31]

India has a long and densely populated coastline of about 7,500 kilometers. Hurricanes, floods, and cyclones are all too common, particularly in states like Odisha. Sea-level increase can lead to death and injury but also reduce fresh-water supply due to saltwater intrusion and contamination through pollutants. Health impacts due to reduced nutrition from agricultural loss and reduced fish catch may also result. Population displacement also leads to vulnerability of people due to improper sanitation, lack of food, and strained health-care centers in the new areas where they migrate.[32]

The concentration of regional air pollutants and aeroallergens will increase with warmer temperatures. This will lead to an increase in respiratory disorders such as asthma, emphysema and chronic bronchitis, and allergy problems. As a result of vehicular pollution, smog will also increase in the cities. Again, this will lead to asthma and respiratory infections in children, chronic obstructive pulmonary disease, and pneumothorax. Air pollution is also likely to increase the incidence of tuberculosis. In cities like Delhi dust storms can exacerbate respiratory problems.

As surface ozone concentrations begin to rise with increasing temperatures the quality of air will deteriorate. Asthma and respiratory and cardiovascular diseases will rise. The government of India is partly addressing the problem in a small way by introducing compressed natural gas (CNG) for transport and replacing wood fire stoves with liquid petroleum gas (LPG) stoves in villages.

Malaria, dengue, and chikungunya are reportedly on the rise in different parts of South Asia. It is generally agreed that malaria will increase in places where temperatures exceed 40 degrees Celsius. Adaptation measures can be further improved by documenting and disseminating the practices of local communities when faced with floods, heat strokes, and mosquito bites.[33]

Ahmedabad is a city prone to high temperatures in summer. A study done by Srinath Reddy and Dileep Mavalankar in the city shows that construction and factory workers, manual labourers, and women working in ill-ventilated kitchens are particularly vulnerable to heat stress. For people in the city, the lakes, parks, trees, and bus stations provide them with some respite from the heat.[34]

Food and the Climate Crisis

> That which, resting in the earth, is other than the earth; which the earth does not know; of which the earth is the body; which controls the earth from within: this is your self, the inner controller, the immortal." (Brhadaranyaka Upanishad III.7.3)[35]

There may be a 10 percent decrease in food-grain production for every one degree rise in temperature, if one goes by the rule of thumb provided by Lester Brown in his book *Plan B*. For a country like India this means a 30 to 40 percent reduction in food-grain production by the end of the century depending on whether the temperature has gone up by 3 or 4 degrees Celsius. A World Bank report on climate change in South Asia paints an equally worrying picture. Here are some extracts from a press release:

> A warming climate will contribute to a slowing in the reduction in poverty. While the lives of everyone in the region will be altered by climate change, the impacts of progressive global warming will fall hardest on the poor who are largely dependent on rain-fed agriculture or live in urban slums. Under 2°C warming by the 2040s, crop production in South Asia may reduce by at least 12%, requiring more than twice the imports to meet per capita demand than is required without climate change. Decreasing food availability can also lead to significant health problems, including childhood stunting, which is projected to increase by 35% by 2050 compared to a scenario without climate change.[36]

The report further states that if temperatures go up by 2 to 2.5 degrees Celsius compared to preindustrial levels, it will have an impact on food security for some 63 million people. This would be due to reduction in water for agricultural production in the river basins of the Indus, Ganges, and Brahmaputra.

The 2014 IPCC report mentions that India will face an agricultural loss of over USD $7 billion by 2030, mainly due to erratic and extreme monsoons. This period is not too far away and it would affect the incomes of 10 percent of the people. A large reduction in wheat yields is also projected in the Indo-Gangetic Plains. By the 2050s South Asia is expected to have a decrease in yield of 16 percent for maize and 11 percent for sorghum. At the same time, the report states that if adaptive strategies are undertaken, 80 percent of the losses could be avoided.[37]

If we have to look for remedial action in this scenario we need to consider some of the problems that beset agriculture in India, largely as a result of government policies. To begin with, the government went all out to support the green revolution in the 1960s, with its package of practices including high-yielding varieties of rice and wheat, chemical fertilizers and pesticides, and credit. Although the green revolution was seen as a spectacular success in India, putting behind the days of famine and mass hunger, only a few people commented on the lack of sustainable practices. Even fewer alerted us to the fact that it took place in places where irrigation was present, and that 60 percent of the cultivated area in India is not irrigated but rain-fed. Out of a total cultivated area in India of 141.0 Mha, rain-fed agriculture represents a whopping 85.0 Mha. About 44 percent of the food-grain production in the country comes from these areas, which includes 75 percent of our pulses (legumes), and 90 percent of our millets and groundnut. In addition, 50 percent of our rural work force makes their livelihood in rain-fed regions and 60 percent of our cattle population is found here. Today, as we learn of the specter of food shortages in the decades ahead, induced by the climate crisis, there are calls for a second green revolution in rain-fed areas, but this time a more sustainable one.

The importance of millets

The campaigns concerning food sovereignty in the context of climate change that Fireflies Ashram and Pipal Tree have been engaging with have mainly centered around millets. As mentioned earlier, millets are grown in rainfed areas, which represent 60 percent of our cultivated area. This is also where the poorest sections of the rural population live.

While millets are well known in India, many other parts of the world people are still ignorant of them; most have not even heard of the term. In South India there is an old Kannada saying which extols the virtues of finger millets, known as ragi among the local people: "The rice eater is weightless like a bird; the one who eats jowar millet is strong like a wolf; and the one who eats ragi is *'nirogi'* [illness free] throughout his life."

There is another well-known poem called "Rama Dhanya Charite," written by Kanakadasa in the 17th century, which talks of the competition between a rice grain and a ragi grain. In the poem a rice grain asserts its superiority over the ragi grain, saying that the latter is low in status and the food of the untouchables. The ragi grain proudly defends itself while affirming that it is proud to be the food of the poor. When the quarrel is taken before Lord Rama, he decides to imprison them both to test their endurance. While the rice grain rots in prison, the ragi grain survives the ordeal and remains fresh.

Culturally, millets are still considered the food of the poor. The whiteness of rice, unfortunately, adds to its status of superiority. Rice is also given a higher status since it is the food of the upper castes while the poor lower castes eat millets. Today, even the poor are slowly giving up millets for high-status rice![38]

Millets are among the oldest cereal grains to be used for domestic consumption in India. They are considered seeded grasses that grow well in dry zones, even in poor soil with little moisture. The main millets cultivated in India are finger millets (ragi), sorghum (known as jowar in Karnataka), pearl millet (sajje), foxtail millet (navane), little millet, kodo millet, proso millet, and barnyard millet. Millets provide superior nutrition compared to rice and wheat. In a nutritionally deficient country like India, millets are three to five times better than rice and wheat in providing proteins, minerals, and vitamins.

There is a lot of malnourishment in the country, particularly among children. Government figures of 2012 indicate that about half the children in the country under five (48 percent) are malnourished. Different sources estimate poverty rates at between 37 and 77 percent of the population, with rural areas having more poverty than cities. Similar indicators find dramatic rates of malnutrition in India (around one fifth of the population), which are especially staggering for children (approximately 48 percent suffer from stunted growth) and women (more than 35 percent have below normal body mass index).[39]

Poor and degraded soil is not a hindrance to cultivating millets. Besides, the soil depth can be less than 15 centimeters. Since millets are usually grown within a mixed farming approach, the soil is considerably enriched with additional rows of pulses (red gram, cowpea, beans, green gram, black gram, lentils) and oilseeds.

Most millet farmers do not use synthetic fertilizers at all. A little farmyard manure will suffice. Therefore, the government can save some money by reducing the use of subsidized synthetic fertilizers. Since millets are also pest resistant, they do not require chemical pesticides. With agricultural emissions representing about 12 percent of total emissions, growing millets can prove to be an effective climate-mitigation strategy as well, since most millet farmers do not use chemical fertilizers and pesticides, which contribute to greenhouse gases. The pollution of groundwater is also reduced by avoiding chemical fertilizers and pesticides.

Growing millets is also a good adaptation strategy for poor farmers since they do not need irrigation. Millets are also much sturdier to withstand fluctuating rainfall and temperatures. Rice and wheat will take a beating in India as climate change kicks in. Millets have a greater capacity to sequester carbon dioxide as they have bigger leaves than rice or wheat. The bigger leaves and larger stalks contribute to increasing fodder stock for cattle. India has a cattle and buffalo population of about 330 million. In 2014, the cattle and buffalo population is likely to rise to 329.7 million head due to growing demand for dairy products.

Unfortunately, millet production and consumption has been going down dramatically. Between 1961 and 2009 there was a sharp decline in the cultivated area of millets: as much as 80 percent for small millets, 46 percent for finger millet, 59 percent for sorghum, and 23 percent for pearl millet. Not only did production come down, but there was a sharp fall in millet consumption as well.[40]

I will mention a few of the reasons for the downswing in millet production and consumption. There is little improvement in the methods of production and the technology needed for threshing and hulling. It is not uncommon to see farmers in rural Karnataka threshing their ragi crop by spreading them on roads and letting cars and trucks run over them. The SRI (System of Rice Intensification) method of rice cultivation that increased rice production is not actively propagated for millets. Crops like BT cotton (i.e., genetically modified cotton for pest resistance) have invaded the areas where millets were cultivated because they are more remunerative to the farmer. In the area in which we work, in H. D. Kote taluk of Mysore district, farmers have largely replaced millets with BT cotton, sugarcane, ginger, bananas, tobacco, and flowers. Many of these crops require huge doses of pesticides and a lot of water, which comes from bore wells.

Government policies have also contributed to diminishing the importance of millets since the Public Distribution System (PDS) largely distributes only rice and wheat. This is a kind of Sanskritization.[41] Sanskritization is an expression coined by the Indian social anthropologist M. N. Srinivas to denote the manner in which lower castes imitate the higher castes in some matters to raise their social status. In this case, the lower castes are imitating the higher castes in their preference for rice and wheat over millets. A Dhan Foundation report states:

> There are no exclusive Government schemes, projects or programs for small millets. Of the available schemes, the most important ones are Initiative for Nutritional Security through Intensive Millets Promotion (INSIMP) and Rainfed Area Development Programme (RADP) as part of Rashtriya Krishi Vikas Yojana (RKVY), and Integrated Cereals Development Programmes in Coarse Cereals based Cropping Systems Areas (ICDP-CC) under Macro Management of Agriculture (MMA). There is a lot of variation across the states on how they utilise these opportunities for promoting millets. Further most of the states usually focus on sorghum, pearl millet and finger millet and leave out many of the small millets while implementing these schemes. Of the schemes mentioned, INSIMP is the only comprehensive initiative to support millets.[42]

Our millet campaign

One of the main problems related to promoting millets and rain-fed agriculture is the lack of support from the government, both central and state. For example, the Public Distribution System (PDS), which distributed rice and wheat to hundreds of millions of poor Indians, did not include millets in its program. But before I go into more detail, let me say a word about PDS.

The implementation of food-security measures by the government of India takes place through the PDS. The purpose of the PDS system is to ensure food security and poverty alleviation. What the PDS does is to identify the poor and needy people, issue them ration cards, procure food grains from farmers, transport the grains to fair-price shops, and sell these grains to those who have received the ration cards at a very low price. The food grains allocated through the PDS system are largely wheat and rice. A few states may include other items like pulses and edible oil.

As far as millets are concerned, there has been a campaign going on for several years led by Millet Network of India (with which Pipal Tree and Fireflies Ashram are collaborating) and other organizations to get the central and state governments to make a commitment to procure and distribute millets in the PDS system. There had been little success until about a year ago, when our advocacy efforts finally paid off.

Although we were working with several hundreds of millet farmers in Mysore district, we appeared to make little headway in getting the state government to look into the plight of small farmers cultivating millets. In July 2013 I accompanied a few of my colleagues from Pipal Tree to meet the local member of Parliament, Mr Dhruv Narayan. We urged him to take up the issue of introducing millets in the PDS at both the state- and central-government levels. He needed no convincing, as farmers in many parts of his constituency grew finger millets (ragi). Together we met with the agriculture minister, whom I happened to know. We were surprised at the overly cordial welcome given us by the agriculture minister, who spent more than two hours discussing millet issues in the state with us. I suspect that the presence of the member of Parliament added to the weight of our discussions. In the course of our meeting the agriculture minister also telephoned the minister of food and civil supplies and senior bureaucrats to apprise them of our quest to get millets included into the PDS in Karnataka.

We did not expect anything to come out of our meeting with the minister of agriculture but we were thoroughly surprised when the chief minister of

Karnataka announced a few days later that the government had decided to introduce millets into the PDS; jowar in the north of Karnataka and ragi in the south. At the moment the cost per kilo through the PDS in Karnataka is one rupee for wheat, rice, and millets (ragi or jowar). P. V. Satheesh from the Millet Network of India, Professor Prakash Kammaradi from the University of Agricultural Sciences, and some millet farmers also joined us in the ongoing campaign. We had several meetings with the minister of food and civil supplies to plead for a higher procurement price for ragi and jowar, so that the millet farmers could break even or make at least a small profit. The minister promised to look into our demands and also asked us to prepare a pilot project to be implemented in a number of villages to experiment with the possibility of a decentralized PDS. Under the pilot project, which is to be implemented in our project area of Mysore district, Stree Shakti, the local women's group, is likely to be responsible for procurement, storage, and distribution of ragi in the villages. If the decentralized PDS becomes successful it would mean that production, procurement, storage, and distribution can take place at the village level. This would save a lot of food miles as far as transportation of food grains is concerned, thus contributing to reducing greenhouse emissions. At the moment rice is transported to Karnataka from the state of Chattisgarh, over many hundreds of kilometers.

The Dance of Krishna

Albert Einstein said that no problem can be solved from the same level of consciousness that created it in the first place.[43] We cannot hope to struggle against climate change with the same consciousness that was responsible for creating it in the first place. Hindu religious leaders echoed Einstein at the Parliament of the World's Religions in Melbourne, Australia, in December 2009:

> Our civilizational crisis cannot be separated from the crisis of our consciousness. Our planet is being polluted because our minds have been sullied by the false consciousness of endless consumption, which in turn leads to the ruthless exploitation of the planet's environment. While some may succeed in reversing the trend through secular ethics many more may need the healing energy derived from religions and spiritualities to create the new consciousness to struggle against climate change.[44]

Let us look at a few Hindu texts that reveal how seriously ancient Hindus have taken the divine nature of the earth. The Brhadaranyaka Upanishad states that the earth is the body of the Divine: "That which, resting in the earth, is other than the earth; which the earth does not know; of which the earth is the body; which controls the earth from within: this is your self, the inner controller, the immortal" (Brhadaranyaka Upanishad III.7.3–23).

Similarly, the Bhagavad Gita states that the divine is present everywhere and in all things: "I am the original fragrance of the earth. I am the taste in water. I am the heat in fire and the sound in space. I am the light of the sun and moon and the life of all that lives" (7.7–9). Hindus have always maintained that the divine pervades the universe, is immanent in all things.

The first verse of Isa Upanishad says: "All this, everything that moves in this moving world/Must be pervaded by the Lord."[45] Everything moves, and the divine moves in everything. Nature is sacred and the divine moves in nature. Vivekananda said that: "every place is the same, every place is the temple of the Lord, every place has become holy . . ." The challenge is to distinguish what is real from the unreal. The *Brihadaranyaka Upanishad* says, "Lead me from the unreal to the real;/lead me from darkness to light;/lead me from death to immortality" (1.3.28). What is illusion and what is real in today's world? In the modern context we have to find a way not to be misled by *maya* or illusion. *Maya* takes us away from our true selves. *Maya* tells us that the goals of life are to become celebrities, to have power, to make a lot of money, to indulge in sex, and so on.

In the consciousness of some people, the world is a commodity to be exploited for profit. They refer to the earth as real estate, resource, mineral, and property. At the altar of money they are prepared to pollute and destroy the life-enhancing potential of planet earth.

Years ago, as a college student, I entered the home of a leading industrialist in Chennai to sell him tickets to the college play. He received us barefoot in his cotton dhoti and shirt. I was surprised by the simplicity of his home, which was free of all ostentation. In traditional India, the value of the simple life was deeply ingrained in the cultural ethos. Human fulfillment was not measured in terms of one's wealth or possessions. The cult of consumerism was totally absent. People recognized the greatness of Gandhi because of his simple life. If he had lived differently he would not have been the mahatma or great soul, and India's nonviolent freedom struggle might have taken a different turn.

We are all interconnected and we live in an interdependent planet. This is what we learn from both the Buddha and Chief Seattle. The pollution emanating from the USA, China, and India will eventually drown a country like Bangladesh, whose ecological footprint is very low. All things are connected, for both good and bad. The story of Indra's net tells us this:

In the heavenly abode of the great god Indra hangs a marvelous and immense net. This vast net opens out in all directions. At each node of the net hangs a single sparkling jewel. Numerous jewels hang in this infinite net. Each jewel glitters away, sparkling like a star, suspended in the net. But if one looks closely, one realizes that each jewel reflects all the other jewels in the net. The process of reflection is infinite.

If each of us is a jewel in Indra's net, then we are all connected with each other. Our dazzle and glitter largely come from the reflection of all the other jewels in the infinite net. We are all influenced and enhanced by each other. If one jewel ceases to sparkle, every other jewel is affected. Likewise, if one jewel glitters more brightly, every jewel glitters more brightly. Every compassionate act affects the whole web. So, if we do something to protect life on the planet, that action influences everybody else on the planet, even if we are not conscious of it. Each action helps to build up the critical mass, which, when reached, can bring the miraculous change we all dream about.

In conclusion, I quote from Ranchor Prime on how gracefully Krishna killed the serpent Kaliya, which was poisoning the river Yamuna:

> Krishna once saved Vrindavan from the threat of pollution. A huge venomous serpent called Kaliya entered the Yamuna river and made his home there. Because of his presence the whole river was poisoned. Trees on the river's banks withered and died. The air itself became so polluted that birds flying over the water fell into it. One day Krishna's cowherd friends drank from the river, not knowing it was poisoned, and fell to the ground unconscious. When Krishna saw this calamity he decided to fight with Kaliya. He dived into the poisonous water and wrestled with the serpent for two hours. During this time the people of Vrindavan watched terrified from the bank. They were sure Krishna would be defeated and were ready to die with him. Suddenly Krishna overcame Kaliya and, climbing on his many heads, began to dance on them in a beautiful way, pressing them down beneath his lotus feet. Exhausted, Kaliya surrendered to Krishna and begged for his mercy. Krishna told him to leave Vrindavan and never return. Once Kaliya had left, the water of the Yamuna river was restored to its natural state and the trees returned to life. Krishna revived his friends and they all celebrated his victory.[46]

Today, Kaliya continues to pollute India and provoke climate change. The faces of Kaliya are many: all the transport vehicles, power plants, and industries that burn fossil fuels, and the toxic chemicals that release emissions into the atmosphere.

Krishna, or the divine force, is present in the millions of people who are consciously trying to live differently, not succumbing to the deceit of consumer society. Krishna is present in the struggles of all the social and environmental activists who wish to save the planet from the perils of climate change.

EPILOGUE

Some Reflections on Ecumenical Action and Advocacy on Climate Change

Ernst Conradie

Introduction

From the perspective of civil society, the World Council of Churches has been one of the most significant role players in addressing global climate change. Given the major efforts from organisations in civil society, alongside the roles of science and government, this is no meagre comment.

There is no need here to document the history of the WCC's involvement in climate change activism, but this is certainly a story that needs to be told at some length.[1] One may point to the very early recognition of the need for sustainability, for example as expressed in the call for a "Just, Participatory and Sustainable Society" at the WCC's Nairobi Assembly in 1975 – long before climate change was on the horizon. In fact, the very term sustainability seems to have been coined at a WCC consultation in Bucharest in 1974. One may also point to many resolutions on climate change adopted since that time, to countless consultations and especially to the ongoing work of the WCC's task team on climate change. There have been several publications dealing directly or indirectly with climate change. This volume may therefore be regarded as the latest in a long lineage of such contributions.

In my reflections on the various contributions included in this volume I will first offer a number of observations on common themes and threads that I found and appreciated in the contributions. I will include some reflections on themes that I found absent from or underplayed in this set of contributions, albeit that they are sometimes implicit. I will then offer a biblical parable on a vision for the future and conclude with some reflections on the triune God's character as a source of inspiration to sustain efforts to stand in solidarity with the victims of climate change.[2]

Common Themes and Threads

There are several common themes and threads that may be identified in the various contributions. These may be noted as follows:

1. As may be expected from a collaborative volume that seeks to make a contribution to ecumenical discourse, there is a welcome diversity in terms of geographical context and confessional traditions represented by the authors of this volume, while there is sufficient diversity in terms of gender and age if not in terms of class and levels of education. There is also a healthy tension between an emphasis on the global and the local, lay and ordained perspectives.

2. The theme of this volume requires a methodological balancing act between social analysis (based on scientific data), theological reflection, theological education, advocacy (policy making), activism, story telling (which seems to be underplayed?) indigenous wisdom, a liturgical spirituality and aesthetic appreciation (epitomized by D'Souza-Krone's welcome contribution). Some contributions focused more on the one than the other, but there is a welcome integration. It is also good to see in a volume of this nature that theological reflection is not dominated by social analysis and a reiteration of data on climate change.

3. In terms of the six core ecumenical themes of unity, mission, Faith and Order, Life and Work, education and worship, there is understandably a

clear tendency to focus on issues of Life and Work. The need for raising an environmental awareness through education is also recognised in several of the essays (the term "education" appears 94 times in the volume, 27 times as "theological education"). Critical reflection on mission is evident in the essays by Meehyun Chung, Werner & Sartorius and Breyer et al (the word "mission" appears 14 times in the volume, 3 times as God's mission). Theological refection on issues related to Faith and Order is evident in several of the essays but tends to be underplayed, except, interestingly enough, in the artistic contribution by D'Souza-Krone.

4. A spectrum of ethical issues related to climate change is addressed. These include a concern over biodiversity (Awad, Behera), deforestation (Andrianos, Motte), desertification (Awad), the production of biofuels such as palm oil (Motte), water management (Adrianos, Awad), food security (Motte, Breyer et al), health (Skuratovskaya, Zhvania & Cornelius), institutional greening of seminaries (Werner) and offices (Breyer et al, Pustelnik), nuclear power (Jea Sophia Oh), energy (Breyer et al), militarism (Meehyun Chung), investments and ecotourism (Behera). There is little emphasis on the ideology of consumerism although issues around the consumption of resources are addressed.

5. Climate change may be regarded as a set of problems that cannot be carefully delineated. It is best to recognise various dimensions of the problem. In this volume this is indeed the case so that the gendered, health, age (see Pustelnik), economic, educational, financial (e.g. Breyer et al) and religious dimensions are all brought into play in one way or the other.

6. The ethical emphasis on justice and sustainability is to be expected in a volume of this nature. The emphasis on peace (Motte), fairness (Breyer et al), reverence for life (Meehyun Chung) and *salim* (enlivening – Jea Sophia Oh) is likewise welcome. There seems to be an interplay between a critique of development discourse and a plea for a form of development qualified by adjectives such as "sustainable," "human" or "social" (see Breyer et al) (the term "development" appears 152 times in the volume). Likewise, there is both a critique of and a plea for responsible "stewardship" (a term that appears 8 times in the volume). My sense is that most contributors recognised the need to maintain a balance between "Ethics," and "Ecclesiology,"

i.e. between an understanding of what the church is and what the church does. However, understandably, an emphasis on ethics is dominant in this volume. The result is that the distinct contribution that churches can make regarding climate change, as compared to other organisations in civil society and faith based organizations, is underplayed. There is a tendency to describe the role of the church merely in functional terms, i.e. in terms of the considerable influence of Christian communities in certain parts of the world.

7. There is a clear emphasis on the sins of greed (Andrianos), but also the sins of pride (arrogance, anthropocentrism – Jea Sophia Oh), alienation (e.g. Behera), moral failure and sloth. The impact of the ideologies of imperialism, capitalism and neo-liberalism (Jea Sophia Oh) is clearly recognised (Meehyun Chung). In Jea Sophia Oh's essay there is a superb critique of the functioning of corporate, technocratic power as opposed to biological sources of energy.

8. In terms of ethical categories, there is an emphasis on the role of visions (together with middle axioms and policies – e.g. Breyer et al), on human rights (Motte, Skuratovskaya, Zhvania & Cornelius), on appropriate values and to a lesser extent on ecological virtues. The virtues of wisdom (the word appears 4 times), humility (Meehyun Chung), courage (the term is used once by by Pustelnik) and resilience (used 3 times) are recognised, as are the categories of sufficiency and restraint (each used once by Breyer et al), but terms such as simplicity or frugality are not used. This is somewhat surprising since the virtue of temperance (in its various manifestations) is clearly crucial for a global response to climate change. The so-called theological virtues of faith (mentioned 6 times), hope (12 times) and love (8 times) are recognized but are not very prominent in this volume. The theme of joy (for some the highest of the theological virtues) is mentioned only once and then with reference to the Korean national anthem!

9. It is also interesting to reflect on the question whether justice is done to the whole work of God. There is an obvious emphasis on God's work of creation, although the term *creation* (which appears 120 times) is used predominantly in the sense of *creatura* and not of *creatio* (except for a discussion in the essay by Breyer et al). One may argue that climate change raises

doctrinal questions primarily about God's providence in terms of conservation (*conservatio*. protecting us from evil), governance of the course of history (*gubernatio*) and cooperation between divine and human agency (*concursus*). Sustainability is a matter of providence if God's mercy endures and sustains us forever (Ps 136). If so, it is striking to note that the words *provide* (used 43 times) and *care* (used 50 times) is used only by D'Souza-Krone with reference to God's provision and care and then primarily in biblical quotations. In all other cases the word *care* is used in the sense of human care for creation. There is one reference to God's protection, but that appears in the Korean national anthem. The term *salvation* is used 16 times and the term *redemption* once, albeit in the form of self-redemption. Several essays recognise the need for social transformation and an inner conversion (*metanoia*) of hearts and minds (e.g. Breyer et al, also Meehyun Chung). There is understandably an emphasis on the church (the term is used 319 times), but very little emphasis on the nature, marks or characteristics of the church, or on God's work of forming, building up or governing the church. Although the word *ministry* appears 17 times this is primarily with reference to the secular government and only three times to specialized ecclesial ministries. There is of course an emphasis on the missions of the church. I could find no reference to God's work of eschatological completion, fulfilment or consummation. The terms God's *kingdom* or God's *reign* appears three times but only once in the sense of God's promise for freedom, justice and reconciliation (see Breyer et al).

10. Since climate change is by definition a global issue that affects people from all nationalities, cultures and faith traditions, it necessarily invites multi-faith collaboration but also inter-religious dialogue. In the various contributions to this volume this need is repeatedly recognised, for example with reference to the Uppsala Interfaith Climate Summit (Grape) and in conversation with Buddhism (Meehyun Chung), Judaism and Islam (Awad). As I will suggest below, such inter-religious dialogue should not remain shallow but should indeed focus on a discernment of God's identity and character – or else what is most important to each faith tradition will be underplayed for the sake of expedience. There is indeed a need to find common ground but the complicity of the Christian tradition in climate change may require from Christians a confession of guilt (and not only a confession of faith), precisely for the sake of multi-faith collaboration.

It is indeed remarkable that the term *faith* appears 52 in this volume, but only 6 times with reference to the Christian faith and 46 times with reference to inter-faith dialogue or faith based organizations. In terms of God's identity, it may be noted that the word *God* is used 94 times in this volume, but never with reference to God as Father or Mother. There are 10 references to Jesus Christ and only two to the Holy Spirit, even though the terms *spirit* or *spirituality* appear 39 times, often in discussions of other religious traditions. This same applies to God's character. How, for example, is an emphasis on God's compassion (Behera) and God's love (a term used twice in the volume) related to and emphasis on God's justice? It is notable that while the word *justice* is used 185 times in this volume it is connected with God's justice only twice. There is no mentioning of God's wrath or God's judgement – which seems to be used mainly be apocalyptic forms of Christianity in a way that is not conducive to climate justice.

The danger here is not merely one of self-secularisation by focusing on the ethical significance of the Christian faith while underplaying its particular content in order to secure possibilities for collaboration. The danger is also that churches may fail to make a contribution to global efforts to address climate change that only they can make. What that contribution could be, remains open for debate. However, given the correlation between historical carbon emissions and those countries where Christianity was established by the advent of the industrial revolution, I would suggest (as hinted above) that an appropriate point of departure may be a confession of guilt (not of faith) on the basis of the message of forgiveness in Jesus Christ.[3] Such a confession may well have to begin within the household of God since some Christians are being victimised by the habits of other Christians.

This begs questions on unity that lie at the core of the ecumenical movement. A volume on ecumenical efforts to address climate change cannot merely celebrate a diversity of perspectives and geographic locations but have to grapple with deeply divisive issues. It seems that the contributors to this volume speak more or less with one voice (which is commendable), albeit from within different geographic contexts, but they belong to churches and live in countries where this is not the case. As long as such tensions are avoided amongst Christians we cannot expect national governments to resolve such issues at any Conference of the Parties. If political leaders cannot commit their countries to take the necessary steps because they are concerned about their relative position of power, church leaders need not do the same. The confession of the unity and catholicity (spread all over the world) of the church, and indeed its holiness and apostolicity, may be the most distinct contribution that churches can make, by confessing their faith to each other and by grasping the implications of that in the context of global climate change! Here is indeed a contribution that churches can make that is not feasible for any other group, including the United Nations.

Missing the Target?

The collection of essays included in this volume should be regarded as a barometer for best practice within the fellowship of churches that constitute the ecumenical movement. The volume is clearly aimed at activism but also at the spiritual sources of inspiration that would guide such activism. Werner and Sartorius draw on the Busan Assembly (2013) of the World Council of Churches to reiterate, "In order to experience radical change, we challenge the churches and other organizations to raise awareness, do and share good practices and engage in advocacy work."

No one should belittle the significance of the ecological reformation of various Christian traditions that is evident in these contributions. Yet, the contributors themselves would be the first to admit that such activism is not commensurate with the scale and scope of global climate change. The question is what actual difference such activism would make to annual carbon emissions globally. Even if such best practice is adopted in each local congregation throughout the fellowship of churches that constitute the World Council of Churches, that may not be sufficient to stabilise the concentration of carbon dioxide in the earth's atmosphere. This is not merely because the 345 member churches of the WCC represent only half a billion Christians worldwide or because less than one third of the global human population are Christains. It is because a reduction of at least 50% in global carbon emissions is required to prevent catastrophic consequences by the end of this century. As is widely recognized, this requires a reduction of up to 80% or 90% in industrialized countries in order to allow some leeway for so-called developing countries to expand their carbon emissions over the medium term. In others words, a cut of 10% or 20% in carbon emissions would not suffice. That is difficult enough as any organizations faced with the need to cut their expenses by 10% would realize.

The scope and scale of what is required may be illustrated by the following two statements by leading scientists (who seem to have become true prophets, going against their own methodological inclinations):

On Earth Day, 22 April 2015 the Earth League (a group of 17 leading scholars) released the *Earth Statement* which urged world leaders to follow up on their commitments to avoid dangerous climate change. This was clearly aimed at influencing the process towards the 21th Conference of the Parties that was held in December 2015 where a new international treaty on climate change was eventually agreed upon.

In the second of eight clauses, the Earth Statement reads:

> The remaining global carbon budget – the limit of what we can still emit in the future – must be well below 1000 Gt CO_2 to have a reasonable chance to hold the 2°C line. Humankind has already emitted around 2000 Gt CO_2 since the beginning of industrialization. Respecting the global carbon budget means leaving at least three quarters of all known fossil fuel reserves in the ground. With current emissions trends, the remaining 1000 Gt CO_2 would be used up within the next 25 years.[4]

The difficulty in this regard is indicated by the group's comment that three quarters of known fossil fuel reserves must be left in the ground if warming was not to breach a rise of 2°C, the safety limit agreed to by the parties in the UN Framework Convention on Climate Change. Even though the Intergovernmental Panel on Climate Change reported in 2014 that a carbon neutral path would cost only 0.06% in reduced global gross domestic product (GDP), it seems that this is a price that governments are unwilling to pay.

A second statement that may help to indicate what is required is the annual report of the Global Carbon Project.[5] The main aim of the project is to gather reliable information on global carbon emissions. A rather telling summary statement of its 2015 Global Carbon Budget (released just before COP 21) reads:

> Carbon dioxide (CO_2) emissions from fossil fuels and industry increased by 0.6% in 2014, with a total of 9.8±0.5 GtC (billion tonnes of carbon) (35.9 $GtCO_2$) emitted to the atmosphere, 60% above 1990 emissions (the Kyoto Protocol reference year). Emissions are projected to decline by -0.6% in 2015 (range -1.6% to +0.5%). In 2014, the ocean and land carbon sinks

> respectively removed 27% and 37% of total CO_2 (fossil fuel and land use change), leaving 36% of emissions in the atmosphere. The ocean sink in 2014 was 2.9±0.5 GtC (10.6 $GtCO_2$), slightly above the 2005-2014 average of 2.6±0.5, and the land sink was 4.1±0.9 GtC (15 $GtCO_2$), the largest sink recorded over the past 60 years, and well above the 3.0±0.8 average of 2005-2014. Total cumulative emissions from 1870 to 2014 were 400±20 GtC (1465 $GtCO_2$) from fossil fuels and cement, and 145±50 GtC (549.6 $GtCO_2$) from land use change. The total of 545±55 GtC was partitioned among the atmosphere (230±5 GtC), ocean (155±20 GtC), and the land (160±60 GtC).

A few observations on this rather technical statement may be helpful. In comparing such figures with the statement by the Earth League above, the difference between using carbon or carbon dioxide as a unit should be noted as well as the degree of uncertainty acknowledged in such measurements. The cumulative carbon emissions are the sum of the total of carbon emitted during a given period of time. Such emissions are either absorbed in the oceans (leading to acidification) or on land (e.g., through photosynthesis) or remain in the atmosphere. The most sobering aspect is that annual carbon emissions have increased quite dramatically since 1990, that emissions are still increasing rapidly (except for 2015) and that predictions over the short to medium term point towards further increases instead of the 50% reduction that is required. The problem is that climate change is related to the cumulative outcome of carbon emissions and the subsequent levels of carbon (dioxide) in the earth's atmosphere. The later annual decreases are effected, the higher the peak for annual emissions will be and the higher the total amount of carbon in the earth's atmosphere will be, even if stabilisation is achieved by 2050 or 2100 – i.e., when further carbon emissions are fully absorbed through photosynthesis. The predicted average temperature increases are correlated with the concentration of carbon / carbon dioxide in the atmosphere. Nevertheless, the slight decline predicted for 2015 suggests that it is indeed possible to decouple economic growth from greenhouse gas emissions through a process of dematerialisation.

Subsequent to these reports, the 21st session of the Conference of the Parties managed to adopt the Paris Agreement on 12 December 2015. This is not the place to assess the significance of this agreement. Suffice it to say that there was clearly a catharsis moment following the failure of several previous attempts to replace the Kyoto Protocol with a binding agreement. Nevertheless, even if governments adhere to their own emissions target, this will at best restrict the predicted rise in average global surface temperature above the pre-industrial era by 2100 to around 2.7 degrees Celsius.[6] This may better than a rise of around 4 degrees Celsius but does not yet address the aim of "less than" 2 degrees Celsius in the Paris Agreement or the more ambitious aim of "not more than" 1.5 degrees Celsius that vulnerable nations campaigned for.

Such a rise in average global surface temperature will have severe implications, especially in areas where a higher rise is probable – and therefore for the global economy. One of the critical factors that will have to be addressed in years to come is that the industrialized global economy would need to refrain from using oil, coal and gas reserves already discovered. What would be required is a transformation of the energy basis of the entire global economy from fossil fuels to sustainable alternatives within a period of 60 years from 1990 to 2050, of which the first two decades were the most important. The Earth League rightly warns that 2015 was our last chance to prevent a rise in average global surface temperature above 2 degrees Celsius. Johan Rockström, the statement's lead author, said: "From a scientific perspective, 2015 is a decisive moment. The window to navigate ourselves free from a 'beyond 2°C future' is barely open. It's the last chance to navigate ourselves towards a desired future."[7]

Given the virtue of hope so deeply embedded in the Christian faith, Christians cannot accept such an assessment as the final word. Apocalyptic scenarios are widely discredited in Christian ecotheology as inducing a paralysing sense of fear.[8] Any prophetic ministry is born from the hope for conversion, that change may yet be possible. I have argued elsewhere[9] for the need to rediscover such hope as a source of inspiration for an environmental praxis, ethos and spirituality. It is hardly possible to motivate people to care for the earth unless they are convinced that there is indeed some future for themselves as earthly creatures. Despair in the face of the environmental crisis will lead to a spirit of resignation. Without hope, environmental activism will lose its impetus and will be fighting a losing battle. The problem is that ecological destruction and economic inequalities are so intractable that this situation elicits unmitigated despair. Despair becomes a self-fulfilling prophecy when it leads people to think that action is futile. Only a persistent hope can therefore energise environmental praxis.

However, there is also an opposite danger here namely that one may seek to standardize the prophetic message along the following lines: God's love extends to the whole household of God. Climate change poses a threat to the well-being of creatures in God's household. Through God's grace the church is called to be an agent of God's mission to bring salvation throughout the earth. The church lives in the hope, based on God's promises, that God's reign will come, on earth as it is in heaven. The problem with this standardized message is not that it is necessarily wrong but that it does not do justice to the more complex biblical roots from which the subsequent Christian tradition(s) draw inspiration. Moreover, it seems to take God's loving response for granted and therefore creates an idol in the name of the God of the Bible. That does not do justice to the dramatic way in which God engages with God's beloved creation in history.

A Prophetic Parable

At the time of writing these reflections (just before Environment Sunday, 2015) I may refer to the reading according to the revised common lectionary of the previous week, namely on Isaiah 6.

This text is well-known for the description of the calling of Isaiah. Verse 3 is often quoted: "Holy, holy, holy is the Lord, the Almighty." The whole earth is filled with God's mighty presence. This refrain is captured in numerous hymns and is often understood to have ecological significance in terms of the grandeur and sanctity of God's whole creation. In contemporary discussions God's might is often underplayed in order to also stress God's vulnerability, but this would not do justice to this particular text.

The second half of the chapter is not equally well-known but this may be where its contemporary ecological significance in the context of climate change is to be found. It describes the deepening disillusionment of the prophet Isaiah in three steps.

The first step is still well-known, namely Isaiah's acknowledgment that he is an unholy man (and an unworthy prophet) living amongst unholy people. Freely contextualized: his carbon footprint is rather high, as is that of the people amongst whom he lives. He has no moral right to speak to others (see verse 5b) about their carbon footprint even if he became aware of the full scope of the impact of climate change. Raising awareness may be the first step but it would not suffice!

The second step of Isaiah's disillusionment lies in the content of the message described in verse 9 and 10. He is called to tell the people that they will not listen to what he has to say. Again freely interpreted: they are to be blinded for the tangible impact of climate change and made deaf to all the prophetic warnings coming from eminent scientists (verse 10a). This is to prevent them from gaining insight and from experiencing God's transformative and healing power (verse 10b). God's message to the prophet is surely ironic and probably meant to have exactly the opposite effect on the audience, but the rhetoric is nevertheless remarkable. At the very least it teaches the prophet to expect a negative response to God's message and to cultivate some resilience.

The third step in Isaiah's disillusionment follows in verse 11 when he asks how long this inability to grasp what is going on will last. Isaiah clearly assumes that his prophetic message will eventually have an effect, but he would not have expected the answer, namely that this will last until the cities are destroyed and the land lies in ruins. Even if only a tenth of the (human population) survives the onslaught and the exile, the land would be even further destroyed.

I have read worst case scenarios[10] where it is predicted that only a tenth of the current human population will survive a 6°C rise in average surface temperature by 2100. Since the population is likely to increase to 9 or 10 billion by 2050, the scale that such a death toll will bring in the last few decades of the 21st century is truly unthinkable. The first of the eight statements in *The Earth Statement* puts this quite graphically: "If we do not act now, there is even a 1 in 10 risk of going beyond 6°C by 2100. We would surely not accept such a high risk of disaster in other realms of society. As a comparison, such a 1 in 10 probability is the equivalent of tolerating about 10,000 airplane crashes every day worldwide!"[11] Moreover, there are some who say that it is already too late to turn the tide and to avert such a crisis, given the cumulative impact of carbon emissions. What this will do to the much vaunted need for economic growth and to the security of financial investments and pension funds is obvious. What may be even more disconcerting is the social instability that is to be expected though the migration of millions of refugees. This will surely present a test for the unity of the church and for solidarity with the victims of climate change. As the WCC noted earlier, there is a distinct possibility that, under difficult circumstances, "love may grow cold" (Mt 24:12). In such a context the church needs to abide by its faith in God alone, in its vibrant hope despite visible success and in its love for the other.[12]

Isaiah's prophetic ministry can be clearly dated during the time of Assyrian and later Babylonian imperialism. In hindsight his apocalyptic imagery came to fulfilment with the fall of Samaria and then Jerusalem and the subsequent exiles of the people of Israel and Judah. In hindsight we also know that the ray of hope that he portrayed in verse 13, namely the stump that will remain when the oak tree is chopped off, also came to fruition. The Christian hope is based on that stump but this does not imply that all future disasters are always averted. One should guard against self-fulfilling prophecies, but prophets who speak truth to power (including ecclesial power) may surely expect a message that is equally disillusioning, disconcerting and disturbing at a time such as this when the God of Life is invoked to lead us to Justice and Peace (Busan 2013).

Reflecting on God's Character

The deeper question is perhaps what God's prophets and priests are called to be doing in such a time when people would not listen. There is no single answer to this question but the biblical narratives give us an array of possibilities emerging from different contexts and periods.

The deepest question, though, is related to the identity and character of the God whose word the prophet is called to bring. Again there is no single answer to this question. The biblical narratives portray a long journey of God's people in which they only gradually and with many deviations from the path came to understand something of the heart of their God. This changed their self-understanding, identity, ethos and praxis.

The royal imagery employed in chapter 6 is awesome and overwhelming. This is a God of immense authority ready to punish the impurities of the nation fiercely and with fearsome consequences. However, this is not the full picture. One also reads about Isaiah's own sins that are forgiven (verse 6), about the possibility of healing (verse 10) and a God looking for an authentic messenger. Indeed, the whole episode is aimed at the long term survival of God's people and God's land, but then on a different basis than the current dispensation.

Elsewhere in the book of Isaiah new insights emerge regarding God's character. A prophet standing in the school of Isaiah learns to see God's immense grace, a soft-spoken God who will not yell in the streets (42:2), the suffering servant who is willing to accept pain on behalf of others (chapter 53).

This discussion leaves one with a pertinent question: what character traits of God are appropriate in the context of climate change? Should one invoke the fierce, almighty, holy One or the vulnerable, broken, despised One? While some may see an irresolvable conflict here, the genius of the Jewish-Christian tradition may well be that such characteristics are held together in the confession of the triune God's love for the world, the whole household of God.

NOTES

Preface

1. Parts of my *Huffington Post* article, "Come into the *Madang*: World Council of Churches and Climate Justice," have been incorporated into this preface; http://www.huffingtonpost.com/grace-jisun-kim/come-into-the-madang-worl_b_5383555.html.

Chapter 1: Doing Theology in a New Way

1. Agenda 21 is a nonbinding, voluntarily implemented action plan of the United Nations with regard to sustainable development. It is a product of the Earth Summit held in Rio de Janeiro, Brazil, in 1992. It is an action agenda for the UN, other multilateral organizations, and individual governments around the world that can be executed at local, national, and global levels. http://www.unep.org/Documents.Multilingual/Default.asp?documentid=52.

2. The Eco-Management and Audit Scheme (EMAS) is a voluntary environmental management instrument, which was developed in 1993 by the European Commission. It enables organizations to assess, manage, and continuously improve their environmental performance. The scheme is globally applicable and open to all types of private and public organizations. In order to register with EMAS, organizations must meet the requirements of the EU EMAS-Regulation.

3. The Church of Sweden has a framework for how it is governed, which stipulates that most decisions be undertaken at congregational levels. This means that the independence of the congregations is rather high. Thus, there can be no national decision that every congregation must have an environment diploma.

4. Diakonia is a faith-based Swedish development organization based in churches other than the Church of Sweden.

5. The Association of World Council of Churches-related Development Organisations in Europe.

6. It is worth mentioning that the Roman Catholic Church is represented and active in the ECEN network.

7. The statement commits the leaders to the following actions: "To inform and inspire people in our own religious and cultural contexts to take responsibility for and to implement effective measures. To challenge political and business leaders where we live and work to develop comprehensive strategies and action. To focus on the struggle against global warming and draw upon our innermost religious convictions about the meaning of life. This commitment is a deeply spiritual question concerning justice, peace and hopes for a future in love and solidarity with all human

beings and the whole of creation." http://www.svenskakyrkan.se/interfaithclimate-summit/inenglish.

8. The letter can be found at http://www.svenskakyrkan.se/biskopsmotet.

Chapter 2: Health as a Human Right

1. Kofi Annan, *We the Peoples: A UN for the 21st Century* (New York: Paradigm, 2014), 223.

2. *The Lancet* 343 (16 May 2009), www.thelancet .com.

3. Round-Up Commission on Human Rights Concludes Fifty-Eighth Session (Economic, Social and Cultural Rights) Appoints Special Reporter on Right to Health (26 April 2002).

4. I wish to gratefully acknowledge the helpful advice of Guillermo Kerber in our composing this chapter.

5. Jorge E. Vinuales, "Legal Techniques for Dealing with Scientific Uncertainty in Environmental Law," *Vanderbilt Journal of Transnational Law* 43, no. 2 (2010): 501.

6. Vandenbergh et al., "Micro-Offsets and Macro-Transformation: An Inconvenient View of Climate Change Justice," *Harvard Environmental Law Review* 33, no. 303 (2009); "Warming of the climate system is unequivocal, as is evident from observations of increases in Global average air and ocean temperatures," from IPCC Climate Change 2007 Synthesis Report 72 (IPCC, 2007); "Supporting Public Awareness Initiatives on the Health Effects on climate Change and Air Pollution," Canadian Public Health Association, http://www.cpha.ca/en/programs/ ccah.aspx

7. World Health Organization, Climate change and health, Fact sheet no. 266, http://www.who.int/mediacentre/factsheets/fs266/en/

8. See https://en.wikipedia.org/wiki/Mikhail_Budyko.

9. Jorge E Vinuales, "Legal Techniques for Dealing with Scientific Uncertainty in Environmental Law," *Vanderbilt Journal of Transnational Law* 43, no. 2 (2010): 501.

10. WHO, Climate change and health, Fact sheet no. 266.

11. Dim Coumou et al., "Quasi-resonant circulation regimes and hemispheric synchronization on extreme weather in boreal summer," *Proceedings of the National Academy of Sciences* 111 no. 34 (2014): 1231–1236, DOI:10.1073/pnas.1412797111.

12. WHO, Climate change and health, Fact sheet no. 266.

13. USGCRP, *Analyses of the Effects of Global Change on Human Health and Welfare and Human Systems*, A Report by the U.S. Climate Change Science Program and the Subcommittee on Global Change Research (Washington, DC: U.S. Environmental Protection Agency, 2008).

14. National Research Council, *Adapting to the Impacts of Climate Change* (Washington, DC: National Academies Press, NRC, 2010).

15. USGCRP, *Analyses of the Effects.*

16. R. Stanwell-Smith, "Water-borne diseases and climate change," *Health Effects of Climate Change in the UK* (Department of Health, England, UK, 2002), 139–51.

17. Ulysses Confalonieri et al., "Human Health," in *Climate Change 2007: Impacts, Adaptation and Vulnerability*, Contribution of Working Group II to the Fourth Assessment Report of the Intergovernmental Panel on Climate Change Party (Cambridge: Cambridge University Press, 2007), 391–431

18. "The Greenhouse Effect and Greenhouse Gases," Windows to the Universe, http://windows2 universe.org/earth/climate/ greenhouse_effect_gases.html.

19. Iain R. Lake et al., "Climate Change and Food Security: Health Impacts in Developed Countries," *Environmental Health Perspectives* 120, no. 11 (2012), DOI:10.1289/ehp.1104424.

20. "What Does Climate Change Have to Do with Health Care?" *Forbes*, 7 April 2013, http://www.forbes.com/sites/skollworldforum/2013/04/07/what-does-climate-change-have-to-do-with-health-care/.

21. L. Skuratovskaya, L. Cornelius, G Kerber Mas, and E. Abramides, "Right to Health for Poverty Reduction and Ecological Justice," *Sustainable Alternatives for Poverty Reduction and Ecological Justice*, Vol. 1, 2nd ed., ed. Lucas Andrianos, Jan Willem Sneep, Guillermo Kerber, and Robin Attfield, 225–231 (Newcastle Upon Tyne: Cambridge Scholars Publishing, 2013).

22. S. Moser and L. Dilling, eds., *Creating a Climate for Change: Communicating Climate Change and Facilitating Social Change* (Cambridge University Press, 2007); Howard Frumkin et al., "Climate Change: The Public Health Response," *American Journal of Public Health* 98, no. 3 (March 2008): 435–45.

23. *Climate change: What you can do?* US Environmental Protection Agency, http://www.epa.gov/climatechange/wycd.

24. http://unfccc.int/gender_and_climate_change/items/7516.php.

25. "What Does Climate Change Have to Do with Health Care?" *Forbes*, 7 April 2013.

26.http://www.undp.org/content/undp/en/home/ourwork/environmentandenergy/strategic_themes/climate_change/focus_areas/gender_and_climatechange/.

27. "Report of the Conference of the Parties on its eighteenth session, held in Doha from 26 November to 8 December 2012," Framework Convention on Climate Change, UN, http://unfccc.int/resource/docs/2012/cop18/eng/08a01.pdf; "Subsidiary Body for Scientific and Technological Advice Thirty-ninth session, Warsaw, 11–16 November 2013," Framework Convention on Climate Change, UN, http://unfccc.int/resource/docs/2013/cop19/eng/04.pdf; "UN Women Launches Year-long Campaign to Build World Where Gender Equality Is Reality," UN News Centre, 22 May 2014, http://www.un.org/apps/news/story.asp?NewsID=47867#.VnsjchFqr-Q

28. Chenyang Xiao-Dayong Hong, "Gender Differences in Environmental Behaviour in China," *Population and Environment* 32, no. 1 (2010): 88–104,

29. Aaron M. Mccright, "The Effect on Climate Change Knowledge and Concern in the American Public," *Population and Environment* 32, no. 1 (2010): 66–87.

30. C. Radel, B. Schmook, and S. McCandless, "Environment, Transnational Labor Migration, and Gender: Case Studies from Southern Yucatán, Mexico and Vermont, USA," *Population and Environment* 32, nos 2-3 (2010): 177–97.

31. Selected Gender Indicators for the Arab Regions and Subregions, 2009, or latest available data from Gender-Responsive Climate Change Adaptation: Ensuring Effectiveness and Sustainability.

32. Skuratovskaya et al., "Right to Health for Poverty Reduction and Ecological Justice."

33. "Brussels G7 Summit Declaration," European Commission, 5 June 2014. On climate change, see paragraphs 10, 11; on health, see, in the part "Development," paras 13, 21, 22, 23.

Chapter 3: Advocacy and Action in Scotland

1. H. Damon Matthews, Tanya L. Graham, Serge Keverian, Cassandra Lamontagne, Donny Seto, and Trevor J Smith, "National Contributions to Observed Global Warming," *Environmental Research Letters* 9 (2014).

2. Museum of Scotland's Shale Oil Industry, http://www.scottishshale.co.uk/index.html.

3. Richard Tipping, "The History of the Scottish Forests Revisited—Part 1 & 2," *Reforesting Scotland* 8 & 9 (1993): 16–21, 18–21.

4. http://www.treesforlife.org.uk/forest/humanimpacts/deforestation.html.

5. W. W. Knox, *A History of the Scottish People. Migration: Scotland's Shifting Population 1840–1940,* www.scran.ac.uk/scotland/pdf/SP2_7migration.pdf.

6. Donald Worster, *A Passion for Nature: The Life of John Muir* (Oxford: Oxford University Press, 2008).

7. "With Scorching Heat and Drought? A Report on the Greenhouse Effect," a series of essays commissioned by the Society, Religion and Technology Project (Edinburgh: St. Andrews Press, 1989).

8. An Environmental Check-up for your Church can be found at http://www.ecocongregationscotland.org/wp-content/uploads/2012/10/Church-Check-Up.pdf.

9. Eco-congregations in England and Wales are managed by A Rocha (see http://ew.ecocongregation.org/); in Germany the *Grune Hahn* ("Green Cockerel") movement is very similar in intent (http://www.kirchliche-dienste.de/projektsites/gruenerhahn-de), and in the USA, GreenFaith is a similar development (http://greenfaith.org/).

10. James S. Pender, *Climate Change, Its Impacts and Possible Community-Based Responses in Bangladesh*, 2d ed. (Dhaka: Church of Bangladeah, 2010).

11. http://www.ecocongregationscotland.org/materials/creation-time/.

12. I am grateful to Margaret Warnock and Rev. Trevor Jamison of Eco-Congregation Scotland for their comments and corrections to the text. Any errors remain my own.

Chapter 4: Youth for Climate Justice

1. EYCE, *EYCE Campaign to Promote Ecological Justice—Concept Paper* (Brussels: EYCE, 2011), 1.

2. This policy paper was adopted by the 39th EYCE General Meeting, which took place in October 2013 in Brno, Czech Republic.

3. EYCE, *Be Eco Heroes! A Manual for EYCE Member Organisations to Become More Sustainable* (Brussels: EYCE, 2013), 2, http://issuu.com/eyce/docs/beecoheroes/5?e=3669107/1522059.

4. This table is based on material in ibid., 4–6.

5. Ibid.

6. Pauline Rosselin, "Be Eco Heroes!" *Leaf* 6 (2014): 15.

7. All the issues of *Leaf* can be found on-line at http://www.eyce.org/activities/campaigns/eyce-campaign-to-promote-ecological-justice-2011-2013/campaign-activities/the-leaf/.

Chapter 5: Environmental Challenges in Finland and in the Lutheran Church

1. "Gratitude, Respect, Moderation: Climate Programme of the Evangelical Lutheran Church of Finland" (Helsinki: Evangelical Lutheran Church of Finland, 2008).

2. Ibid., 52.

3. Ibid., 55.

4. The original Finnish text is titled "Kiitollisuus, kunnioitus, kohtuuus. Suomen evenkelisluterilaisen kirkon ilmasto-ohjelma" (Helsinki, 2008). The Swedish version is titled "Tacksamhet, respekt, måttfullhet. Klimatprogram för Evagelisk-lutherska kyrkan i Finland" (Helsingfors, 2008).

5. Finnish web pages of the Eco-Fast campaign: http://www.ekopaasto.fi/fi/etusivu.

6. The opening speech in English may be found at: https://arkkihiippakunta-fi-d.aldone.fi/arkkipiispa/puheet-ja-kirjoitukset/puheenvuorot-ja-tervehdykset/2013/opening-address-at-the-general-s2/.

7. Ibid.

8. Ibid.

9. Ibid.

10. Ibid.

Chapter 6: Advocacy for Eco-Justice in Palestine

1. Climate Change Adaptation Strategy and Programme of Action for the Palestinian Authority, PEQA and UNDP, 2010. Special thanks to Kaitlyn Baldrige and Sheldon Way for their assistance and efforts in this chapter.

2. Ussama Abu Ali, et al., *Field Guide in Food Security* (Jerusalem: Ma'an Development Center, 2006).

3. Climate Change Adaptation Strategy and Programme of Action for the Palestinian Authority, PEQA and UNDP, June 2011.

4. Malcolm Lowe, "Palestinian Water (and Martin Schulz): A Lack of Logic," Gatestone Institute, 3 March 2014, http://www.gatestoneinstitute.org/4198/israel-palestinian-water-martin-schulz.

5. "Gaza in 2020: A Liveable Place?," 28 August 2012, United Nations Relief and Works Agency, http://www.unrwa.org/newsroom/press-releases/gaza-2020-liveable-place.

6. Dr Yousef Abu Safieh, "The United Nations Environmental Program (UNEP) Reports on Palestine," presented at The Environmental Education Center's Third Palestinian Environmental Awareness and Education Conference, entitled "Eco-Justice of Palestine," December 2012.

7. Simon Awad, "Birds of Palestine: A decade of ringing and census data" (East Jerusalem: Environmental Education Center, 2009).

8. See the Environmental Education Center's website at http://www.eecp.org.

9. "Uppsala Interfaith Climate Manifesto" (Uppsala: Church of Sweden, 2008), https://www.svenskakyrkan.se/default.aspx?id=664984.

10. The Environmental Education Center's Third Palestinian Environmental Awareness and Education Conference entitled "Eco-Justice of Palestine," December 2012.

11. Dr Guillermo Kerber, "Eco-Justice in Palestine-Theological and Ethical Insights," address at the Environmental Education Center's Third Palestinian Environmental Awareness and Education Conference entitled "Eco-Justice of Palestine," December 2012.

12. Bishop Dr Munib Younan, Introduction to The Environmental Education Center's Third Palestinian Environmental Awareness and Education Conference entitled "Eco-Justice of Palestine," December 2012.

13. Bishop Dr Munib Younan, "Like Trees Planted by Streams of Water," address to the meeting of the Lutheran World Federation (LWF) Council in Indonesia, June 2014.

14. See http://www.elcjhl.org.

Chapter 7: One Earth, One Sea, One Sky

1. This article is taken from a presentation prepared for a meeting of the World Council of Churches' Climate Change Working Group in Bossey, Switzerland, May 2012.

2. Aldo Leopold, *A Sand County Almanac* (New York: Oxford University Press, 1949), 225; as quoted in Ross L. Smillie, *Practicing Reverence: An Ethic for Sustainable Earth Communities* (Kelowna, BC: CopperHouse/Wood Lake, 2011), 67.

3. Walter Brueggemann, *The Land: Place as Gift, Promise, and Challenge in Biblical Faith*, rev. ed., Overtures to Biblical Theology (Minneapolis: Fortress Press, 2002), 2–3.

4. Alanna Mitchell, *Sea Sick: The Hidden Crisis in the Global Ocean* (Toronto: McClelland & Stewart, 2008), 12. More resources may be found on her website, http://www.alannamitchell.com/.

5. Melanie Fitzpatrick, "The New 400ppm World: CO_2 Measurements at Mauna Loa Continue to Climb," blog post on the website of the Union of Concerned Scientists, 20 March 2014, http://blog.ucsusa.org/400ppm-co2-mauna-loa-455.

6. Figures are from the website of the National Organization and Atmospheric Administration (NOAA) of the USA, http://www.esrl.noaa.gov/gmd/; especially the section on Atmospheric CO_2—Weekly Data Mauna Loa Observatory | NOAA-ESRL Data, http://www.esrl.noaa.gov/gmd/ccgg/trends/.

7. In chemistry, pH is a measure of the activity of the (solvated) hydrogen ion; it is measured on a scale from 1 to 14, with 1 being very acidic and 14 very alkaline or basic. Pure water has a pH very close to 7 at 25 degrees C.

8. Magdalena A. Balmaseda, Kevin E. Trenberth, and Erland Källén, "Distinctive Climate Signals in Reanalysis of Global Ocean Heat Content," *Geophysical Research Letters*, 10 May 2013, http://onlinelibrary.wiley.com/doi/10.1002/grl.50382/abstract.

9. Global Ocean Commission, "Life in the High Seas storing 5—million tonnes of atmospheric carbon every year," 5 June 2014, http://www.globaloceancommission.org/news/life-in-the-high-seas-storing-500-million-tonnes-of-atmospheric-carbon-every-year/?utm_campaign=2014-06-05%20Latest.html&utm_medium=email&utm_source=EThe

10. The United Church of Canada, "A Song of Faith: A Statement of Faith of the United Church of Canada," August 2006, http://www.united-church.ca/beliefs/statements/songfaith.

11. Walter Wink, *Engaging the Powers: Discernment and Resistance in a World of Domination* (Minneapolis: Fortress Press, 1992), 56.

12. World Council of Churches, "Statement on Just Economy for Sustainable Life" (Geneva, 2010).

13. World Council of Churches, "Statement on Eco-justice and Ecological Debt" (Geneva, 2010).

14. Barbara Rossing, from transcript of remarks at WCC North America Poverty, Wealth & Ecology Forum and Hearings, Calgary, Alberta, November 2011.

15. A personal reflection from Keith Helmuth to colleagues of the Canadian Council of Churches' Commission on Justice and Peace, 2013.

16. The United Church of Canada, *A Song of Faith*, http://www.united-church.ca/beliefs/statements/songfaith.

17. Smillie, *Practicing Reverence*, 129, 186.

Chapter 8: Religious Responses to Climate Change

1. Nicene-Constantinopolitan Creed, adopted at the Second Ecumenical Council, 381 CE.

2. The Universal Declaration of Human Rights, adopted by the United Nations General Assembly on 10 December 1948 at the Palais de Chaillot, Paris, France, www.un.org/en/documents/udhr/.

3. UNCED, Rio de Janeiro, Brazil, June 1992.

4. IPCC, 2013: Summary for Policymakers. In *Climate Change 2013: The Physical Science Basis. Contribution of Working Group I to the Fifth Assessment Report of the Intergovernmental Panel on Climate Change* ed. T. F. Stocker, D. Qin, G.-K. Plattner, M. Tignor, S.K. Allen, J. Boschung, A. Nauels, Y. Xia, V. Bex and P.M. Midgley. (Cambridge University Press, Cambridge, United Kingdom and New York, NY, USA 2013), www.climatechange2013.org/images/report/WG1AR5_SPM_FINAL.pdf; see p. 27.

5. World Council of Churches, "Justice and peace must prevail: We must not betray life!," Statement from the World Council of Churches to the High-Level Ministerial Segment of the 19th Session of the Conference of the Parties–COP19 to the UNFCCC 9th Session of the Meeting of the Parties–CMP9 to the Kyoto Protocol, Friday, 22 November 2013, www.unfccc.int/files/meetings/warsaw_nov_2013/statements/application/pdf/cop19_hls_faith.pdf.

6. Ecumenical Patriarch Dimitrios, "Encyclical Letter on the Environment, 1st September 1989"; The Ecumenical Patriarchate and the World Wide Fund for Nature (WWF): "Orthodoxy and the Ecological Crisis," Phanar, Istanbul, Turkey, and Gland, Switzerland, 1990.

7. Ecumenical Patriarch Bartholomew, on the Day of Prayer for the Protection of All Creation, on 1 September 1994.

8. Ecumenical Patriarch Bartholomew, "Address at the Environmental Symposium," Santa Barbara, California, USA, November 1997.

9. Ecumenical Patriarch Bartholomew, "Statement for the WCC Working Group on Water," August 2005 (Geneva: World Council of Churches, January 2006).

10. Ecumenical Patriarch Bartholomew: "Statement for the WCC Working Group on Climate Change," August 2005. (Geneva: WCC, January 2006).

11. Message of Ecumenical Patriarch Bartholomew on the Day for the Protection of the Environment, 1 September 2010.

12. H.H. The Fourteenth Dalai Lama of Tibet, *My Tibet* (London: Thames and Hudson, 1990), 79–80.

13. "The Buddhist Attitude to Animal Life," in Venerable K. Sri Dhammananda, "What Buddhists Believe," http://www.budsas.org/ebud/whatbudbeliev/170.htm.

14. Narrated in Ahmad ibn Hanbal, Musnad, IV, 61, 374.

15. Fazlun Khalid and Ali Kh. Thani, "Teachers Guide Book for Islamic Environmental Education" (Birmingham, UK: The Islamic Foundation for Ecology and Environmental Sciences, 2007), http://www.ifees.org.uk/wp-content/uploads/2015/04/13751866541.pdf.

16. Dr Karan Singh, *Essays on Hinduism* (Columbia, MO: South Asia Books, 1999).

17. WCC 10th Assembly, Busan, Republic of Korea, 30 October–8 November 2013.

18. UN and Climate Change, "Climate Summit 2014," http://www.un.org/climatechange/climate-summit-2014.

19. Climate Summit 2014, "2014 Climate Change Summary–Chair's Summary," http://www.un.org/climatechange/summit/2014/09/2014-climate-change-summary chairs-summary/

20. Interfaith Summit on Climate Change, http://www.interfaithclimate.org.

21. "Message by His All-Holiness Ecumenical Patriarch Bartholomew to the Interfaith Summit on Climate Change," New York, 21–22 September 2014, http://www.oikoumene.org/en/press-centre/files/PatriarchmessageInterfaithClimateSummit.pdf.

22. Interfaith Summit on Climate Change, "The Statement," http://www.interfaithclimate.org/the-statement.

Chapter 10: From Climate Justice to the *Energiewende*

1. This essay is written by staff members of the Evangelical Church in the Rhineland (http://www.ekir.de) and the Evangelical Church of Westphalia (http://www.evangelisch-in-westfalen.de). Most of them also participated at the annual meeting 2014 of the Climate Change Working Group of the WCC in Wuppertal (Germany), where this book project was developed. Opinions expressed in this essay reflect the views of the authors.

The Evangelical Church of Westphalia (EKvW) is intensively working on climate protection and climate justice. Consultants from the Institute for Church and Society of the EKvW and the Institute for Mission, Ecumenical Affairs and Global Responsibility of the EKvW contributed to the essay.

We thank Gabriele Walz from the Institute for Mission, Ecumenical Affairs and Global Responsibility of the EKvW for her engaged support with our translation and Anne Freudenberg from the Center for Mission and Ecumenism of the Evangelical Lutheran Church in Northern Germany for her remarks.

2. See "World in Transition—A Social Contract for Sustainability," Flagship Report 2011; http://www.wbgu.de/en/flagship-reports/.

3. Many concrete examples originate from the North Rhine-Westphalian context. This is owed to the fact that the authors of this article are employees of the Evangelical Church in the Rhineland and the Evangelical Church of Westphalia.

4. Declaration of the 11th Synod of the Evangelical Church in Germany, "'There is enough for everyone'—World Food Security and Sustainable Agriculture, " Düsseldorf 2013; http://www.ekd.de/english/5020.html.

5. "'Turning to life'—Sustainable Development in the Context of Climate Change," A Memorandum of the Council of the Evangelical Church in Germany, Berlin/Hannover, May 2009; http://www.ekd.de/english/download/memorandum_turning_to_life.pdf.

6. Ibid., 12.

7. Declaration of the 11th Synod of the Evangelical Church in Germany, "'There is enough for everyone'—World Food Security and Sustainable Agriculture," Düsseldorf 2013, http://www.ekd.de/english/5020.html.

8. Further information can be found on http://www.kirchliches-umweltmanagement.de.

9. http://www.klima-kollekte.de.

10. http://www.zukunft-einkaufen.de.

11. http://www.klimagerechtigkeit.de/.

12. http://www.klimawandel-lebenswandel.de/.

13. http://www.gutesleben-fueralle.de/.

14. http://www.poweredbyheaven.de/.

15. http://www.ekd.de/download/ekd_texte_113.pdf.

16. http://www.brot-fuer-die-welt.de/en/bread-for-the-world.html.

17. http://www.aprodev.eu/.

18. http://www.climate-alliance-germany.de/.

19. The English version of the platforms guideline can be found at http://www.evangelisch-in-westfalen.de/fileadmin/ekvw/dokumente/projekte/klimaplattform/climate_of_justice_screen.pdf.

20. http://www.klimapilgern.de.

21. Minute on Climate Justice in Busan, referring to Deut. 10:17-18.

22. See http://www.climateactionday.org.

23. See http://www.oekumene-ack.de/themen/glaubenspraxis/oekumenischer-tag-der-schoepfung/.

Chapter 11: Deforestation, Poverty, and Climate Change

1. World Commission on Environment and Development (WCED), *Our Common Future* (Oxford: Oxford University Press, 1987).

2. Lester R. Brown, Christopher Flavin, and Hilary F. French, *State of the World 2000: A Worldwatch Institute Report on Progress Towards a Sustainable Society* (New York: Norton, 2000).

3. International Union for the Conservation of Nature, United Nations Environment Program, and WorldWide Fund for Nature (IUCN/UNEP/WWF), *Caring for the Earth: A Strategy for Sustainable Living* (Gland, Switzerland: IUCN, 1991).

4. E. O. Wilson, *The Future of Life* (New York: Vintage, 2002).

5. Ron Nielsen, *The Little Green Handbook: Seven Trends Shaping the Future of Our Planet* (New York: Picador, 2006).

6. H. J. Zimmermann, *Fuzzy Set Theory and Its Applications* (Boston: Kluwer, 1991).

7. Yannis Phillis and Luc A. Andriantiatsaholiniaina, "Sustainability: An Ill-Defined Concept and Its Assessment Using Fuzzy Logic," *Ecological Economics* 37, no. 3 (2001): 435–56.

8. Luc Andriantiatsaholiniaina et al., "Evaluating Strategies for Sustainable Development: Fuzzy Logic Reasoning and Sensitivity Analysis," *Ecological Economics* 48, no. 2 (2001): 149–72.

9. Lucas Andrianos, *Multidimensional Structural Greed index (MSGI): A prerequisite tool for sustainable development policy and eco-justice* (Geneva: WCC Publications, 2013).

10. www.oac.gr.

11. WCED, *Our Common Future.*

12. Luc A. Andriantiatsaholiniaina, *Sustainability Assessment using Fuzzy Evaluation*, Ph.D. thesis, Chania, Technical University of Crete, 2001.

13. Organization for Economic Co-operation and Development (OECD), *Environmental Indicators: A Preliminary Set* (Paris: OECD, 1991).

14. Andriantiatsaholiniaina, et al., "Evaluating Strategies."

15. M. A. Hersh, "Sustainable Decision Making: The role of decision support systems," *IEEE Transactions on Systems, Man, and Cybernetics Part C (Applications and Review)* 29, no. 3 (1999): 395–408.

16. Andriantiatsaholiniaina et al., "Evaluating Strategies."

17. Heli Lu, Xi Wang, Yan Zhang, Weiyang Yan, and Jinping Zhang, "Modelling Forest Fragmentation and Carbon Emissions for REDD plus" (The Second SREE Conference on Engineering Modelling and Simulation [CEMS]), *Procedia Engineering* 37 (2012): 333–38,

18. Mshana Rogate, ed., *Linking Poverty, Wealth and Ecology in Africa* (Geneva: WCC Publications, 2012).

19. Lucas Andrianos, "Setting the Greed Line Using Biblical Insights and Sustainability Ethics in order to Avert Ecological Crisis," in *Ecological Theology and Environmental Ethics*, vol. 2 (ECOTHEE-11) (Chania, Crete: OAC Publications, 2012), 190–202.

20. Konrad Raiser, "Theological and Ethical Considerations regarding Wealth and the Call for Establishing a Greed Line," *Ecumenical Review* 63, no. 3 (2011): 278–94.

21. Athena Peralta, ed., *Poverty, Wealth and Ecology in Asia and the Pacific: Ecumenical Perspectives* (Geneva/Chiang Mai/Suva: WCC/CCA/PCC Publications, 2010), http://www.cca.org.hk/resource/books/olbooks/poverty_wealth_and_ecology_in_ap.pdf.

22. Larrea Carlos, *Inequality, Sustainability and the Greed Line: A Conceptual and Empirical Approach*, Ecumenical Review Vol. 63, Issue 3 (2011).

23. Lucas Andrianos et al., ed., *Sustainable Alternatives for Poverty Reduction and Eco-Justice Book of Proceedings* (SAPREJ-12), vol. 1, (Chania, Crete: OAC Publications, 2013).

24. Grady J. Harper, Marc Steininger, Compton Tucker, Daniel Juhn, and Frank Hawkins, "Fifty Years of Deforestation and Forest Fragmentation in Madagascar, " *Environmental Conservation* 34, no. 4 (2007): 325–33.

25. Lucas Andrianos et al., eds., *Ecological Theology and Environmental Ethics*, ECOTHEE-I (Chania, Crete: OAC Publications, 2009).

26. Lucas Andrianos et al., eds., *Conservation and Sustainable Use of Wild Plant Diversity*, CSUWPD (Chania, Crete: OAC Publications, 2011).

Chapter 12: Eco-Theology, Climate Justice, and Theological Education

1. See https://institute.oikoumene.org/en/study-at-bossey/continuing-formation/green-churches-ecology-theology-and-justice-in-practice.

2. See the work of Prof. Ernst Conradie, at http://www.uwc.ac.za/Biography/Pages/Ernst-M-Conradie.aspx.

3. GlobTheoLib, "Global Survey on Theological Education: Summary of Main Findings," WCC 10th Assembly, Busan, South Korea, 30 October to 8 November 2013, http://www.globethics.net/web/gtl/research/global-survey.

4. See section 6 in ibid.

5. Further details and documents on the international study report on theological education, including first commentary papers from various regions of the world on the findings, are included in the website folder on ecumenical conversation 6 at the Busan assembly; http://www.globethics.net/web/ecumenical-theological-education/documents?layoutPlid=13227594.

6. For details, see http://www.greenseminaries.org/index.php/aboutus.

7. All information taken from the final report, 9/14/12, http://www.greenseminaries.org/images/PDF_Files/Survey_Report_for_Publication_final_formatted.pdf.

8. Only a selection is listed, for details see the full report (ibid.).

9. http://greenfaith.org/programs/certification. Special gratitude is expressed by the authors to Fletcher Harper, executive director of GreenFaith, who updated this article with valuable informaion.

10. See the Global Survey on Ecotheology, Climate Justice and Food Security: http://www.globethics.net/web/gtl/ecotheology.

11. While some substantial mapping of resources has been done by some outstanding scholars and research centres, (e.g., see bibliographies on eco-theology from Prof. Ernst Conradie in: Conradie, *Christianity and Ecological Theology: Resources for Further Research*, Study Guides in Religion and Theology 11 [Stellenbosch: SUN Press, 2006]; and idem, *Christianity and Earthkeeping: In Search of an Inspiring Vision*, Resources in Religion and Theology 16 [Stellenbosch: SUN Press, 2011]), we have not developed a survey particularly on resources and best-practice models for training in eco-theology, climate justice, and food issues.

12. http://www.globethics.net/web/gtl/ecotheology/take-the-survey.

13. http://www.globethics.net/web/gtl/ecotheology.

14. Such as Michael Northcott's *A Moral Climate: The Ethics of Global Warming* (Maryknoll, NY: Orbis, 2007), or contributions from Sallie McFague, Anne Primavesi, and Ernst Conradie. See also the standard bibliography from WCC-ETE for teaching Social Ethics and Ecumenism, section on creation.

15. The formal ETCF project owner is the GlobeTheoLib in Geneva, which is part of the Globethics.net Foundation. This research project will be accompanied by a steering group, which is composed of Dr Dietrich Werner (senior theological advisor, Bread for the World, Berlin) and Dr Stephen Brown (program executive, Globethics.net/Global Digital Library for Theology and Ecumenism, Geneva), Dr Guillermo Kerber (WCC executive staff for care for creation and climate justice, Geneva), and Rev. Andar Parlindungan (executive secretary for training and empowerment, United Evangelical Mission, Wuppertal).

16. See John Creasy, "Ecotheology and the Role of the Church," Pittsburgh Theological Seminary, at http://www.pts.edu/blog/ecotheology-church-earth-day/.

Chapter 13: *Salim*, Ever-Recycling Energy

1. "Obama Administration Announces Loan Guarantees to Construct New Nuclear Power Reactors in Georgia," Office of Nuclear Energy, 16 February 2010, http://energy.gov/ne/articles/obama-administration-announces-loan-guarantees-construct-new-nuclear.

2. Jea Sophia Oh, *A Postcolonial Theology of Life: Planetarity East and West* (Upland, CA: Sopher Press, 2011), 178.

3. Jea Sophia Oh, "Delivering Peace out of the Broken Womb: A Postcolonial Interreligious Perspective," *Current Dialogue* (Special Issue for the 10th Assembly of the World Council of Churches) 55 (September 2013): 27.

4. Climate Change and President Obama's Action Plan, *The White House* (23 September, 2014).

5. Robert J. C. Young, *Postcolonialism: A Historical Introduction* (Oxford: Blackwell, 2001), 27.

6. Catherine Keller, *God and Power: Counter-Apocalyptic Journeys* (Minneapolis: Fortress Press, 2005), 123.

7. "US Nuclear Power Policy," *World Nuclear Association* (December 2014), http://www.world-nuclear.org/info/Country-Profiles/Countries-T-Z/USA--Nuclear-Power-Policy/.

8. Staff Writers, "Seoul gives green light to two nuclear reactors," *Nuclear Power Daily*, Seoul (UPI), 29 January 2013, http://www.nuclearpowerdaily.com/reports/Seoul_gives_green_light_to_two_nuclear_reactors_999.html.

9. Park Ju-Min and Cho Mee-Young, "South Korea cancer victims bring class action against nuclear operator," *Reuters*, 16 December 2014, http://www.reuters.com/article/2014/12/16/us-southkorea-nuclear-cancer-idUSKBN0JU0OZ20141216.

10. Kazuaki Nagata, "Fukushima puts East Asia nuclear policies on notice," *Japan Times*, 1 February 2012, http://www.japantimes.co.jp/news/2012/02/01/national/fukushima-puts-east-asia-nuclear-policies-on-notice/#.ViVVw6SB__5.

11. Simon Mundy, "South Korean cuts target for nuclear power," *Financial Times*, 14 January 2014, http://www.ft.com/cms/s/0/4e8c1872-7cf7-11e3-81dd-00144feabdc0.html#axzz3p355N91G.

12. Alexander Glaser, "After the nuclear renaissance: The age of discovery," *Bulletin of Atomic Scientists,* 17 March 2011, http://thebulletin.org/after-nuclear-renaissance-age-discovery.

13. Jung Gyung-Lan, "'We want a nuclear-free peaceful world,' say South Korea's Women," *WNN—Women's New Network*, 18 January 2012, http://womennewsnetwork.net/2012/01/18/we-want-a-nuclear-free-world/.

14. Oh, *Postcolonial Theology of Life*, 175.

15. Vandana Shiva, *Making Peace with the Earth* (New York: Pluto Press, 2012), 5.

16. Oh, *Postcolonial Theology of Life*, 2, 61–82.

17. Shiva, *Making Peace with the Earth*, 73–74.

18. Vandana Shiva, *Soil Not Oil: Environmental Justice in a Time of Climate Crisis* (Cambridge, MA: South End Press, 2008), 129.

19 Gayatri Chakravorty Spivak, *Outside in the Teaching Machine* (New York: Routledge, 1993), 3, 172.

20 Gilles Deleuze, "On Capitalism and Desire," in *Desert Islands and Other Texts*, trans. Michael Taormia (New York: Semiotext, 2004), 263.

21 Ibid., 267.

22. Shiva, *Making Peace with the Earth*, 11.

23. John B. Cobb Jr, *Sustainability, Economics, Ecology, and Justice* (Maryknoll, NY: Orbis, 1992).

24. A. J. Peterson, "Operational Experience of a Residential Photovoltaic Hybrid System," *Solar Energy* 65, no. 4 (1999): 227–35.

25. Michel Foucault, *The History of Sexuality, vol. 1: An Introduction*, trans. Robert Hurley (New York: Vintage, 1978), 94–95.

26. Homi K. Bhabha, *The Location of Culture* (New York: Routledge, 1994), 112.

27. Ibid., 114.

28. Oh, *Postcolonial Theology of Life*, 82.

29. Shiva, *Soil Not Oil*, 136.

30. Whitney A. Bauman, *Religion and Ecology: Developing a Planetary Ethics* (New York: Columbia University Press, 2014), 172.

31. Grace Ji-Sun Kim, *Colonialism, Han, and the Transformative Spirit* (New York: Palgrave MacMillan, 2013), 69.

32. Shiva, *Soil Not Oil*, 144.

33. Deane Curtin, *Environmental Ethics for a Postcolonial World* (New York: Rowman and Littlefield, 2005), 25.

34. Willis Jenkins, *The Future of Ethics: Sustainability, Social Justice, and Religious Creativity* (Washington, DC: Georgetown University Press, 2013), 240.

35. Larry Rasmussen, *Earth-Honoring Faith: Religious Ethics in a New Key* (New York: Oxford University Press, 2013), 114.

36. The Grand Canal Project got underway in 2008 by the Lee Myung-Bak government as the so-called Korean Green New Deal Project. In the name of economic progress, this project has destroyed the beauty of nature and threatened the circle of life. In order to carve out the waterway, numerous trees have been cut down.

37. In January 2011, the South Korean Navy began construction on a USD $970 million base in Gangjeong town, Jeju. Once completed, it will be home to 20 warships, including submarines, that the navy says will protect shipping lanes for South Korea's export-driven economy, which is dependent on imported oil. On 7 March 2012, the South Korean Navy, together with the construction company Samsung Corporation, started blasting out the Gurumbi Rock foundations in the coastline. This project has greatly damaged the ecosystem of the coastline.

38. When the International Monetary Fund (IMF) and South Korea negotiated the largest IMF rescue package ever in early December 1997, of approximately USD $57 billion, it came as a shock to Koreans. The situation became popularly known as the "IMF Crisis." Confronting the "IMF crisis," Korean women started the "*Ah-na-ba-da* Movement" as one of the *salim* movements to recover the Korean economy. Oh, *Postcolonial Theology of Life*, 36.

39. Chung Hyun Kung, *Miraeeseo On Pyeonji*: *Goddess-spell According to Hyun Kyung* [A Letter from the Future] (Seoul: Yeolimwon, 2001), 236–40.

40. Ibid., 18.

41. Alfred North Whitehead, *The Function of Reason* (Princeton: Princeton University Press, 1929), 8.

42. Oh, *Postcolonial Theology of Life*, 33.

Chapter 14: Rice Is Heaven

1. Lynn White Jr, "The Historical Roots of Our Ecologic Crisis," *Science* 155, no. 3767 (10 March 1967): 1203–1207.

2. Lynn White Jr, *Medieval Technology and Social Change* (London: Oxford University Press, 1964), 56.

3. For example, in Korean construction projects there was a regulation that if there was a tree in the yard, then that tree should not be harmed but a way should be found to protect it.

4. Cf. http://www.kdi.re.kr/infor/ep_view.jsp?num=98021.

5. But in small-scale movements, with their awareness training, there is a significant tendency to do something against such profit-oriented thinking and acting. Examples include pastors in rural areas who provide environmental education and engage in eco-agriculture, solar panels on church roofs or buildings, or the movement to reduce food waste. There are also various forms of network training under multi-religious groups to increase profit awareness and to unite for action. There is another positive example of a successful movement for protecting the land from destruction during development projects. Keyangsan in Incheon is a favorite mountain for Incheon citizens, where many people can observe the process of the seasons and promote awareness for becoming a friend with nature. However, it is also an attractive location for a jaebul (a Korean form of business conglomerate) to develop golf courses for the minority of rich people.

A protection movement against this plan was started by a female environmental activist, Shin Jeong-Eun, who stayed on a tree for 56 days. It was a battle between

David and Goliath. This was continued by a Protestant pastor, Yoon In-Jung, who continued this action for another 155 days, an action that was supported by his congregation. Young or old, male or female, Christian or non-Christian did not matter, as belonging to this movement sparked and attracted the attention of ordinary citizens. As a result, the city government officially made a resolution not to permit the building of a leisure park for golf on behalf of the rich minority, but to protect it for the poor majority. This is a very constructive case that could be multiplied elsewhere. For more about this demonstration see Yoon In-Jung, Shin Jeong-Eun, *Letter from Pineforest: 210 days on the Keyang Mountain* (Seoul: Dongyeon, 2008), http://www.greenkorea.org.

6. KBS documentary report, "Saying hello to 4 rivers," http://news.kbs.co.kr/news/NewsView.do?SEARCH_NEWS_CODE=2914476.

7. Among the articles in SOFA 3, there are two items that refer to the fact that the U.S. government has an environmental responsibility to protect nature and human health.

8. Cf. http://www.greenkorea.org.

9. Cf. Keun-Joo Christine Pae, "Feminist Activism as Interfaith Dialogue: A Lesson from Gangjeong Village of Jeju Island, Korea," *Journal of Korean Religions* 5, no. 1 (April 2014): 55–69.

10. Due to the high cost of reunification, the majority of young people do not have much interest in reunification. They prefer the cheap option of the status quo. This means living in a divided country and dualistic hostility.

11. This part is based on the following article but is slightly changed: Meehyun Chung, "Seeking the Lost Threefold Thoughts: Relationships with God, Earth and Human Beings," *Madang, International Journal of Contextual Theology in East Asia* 22 (2014): 115–36.

12. This kind of Korean triune concept could support an understanding of the profound mystery of the triune God as *vestigia trinitatis*.

13. This heritage of Christianity did not become mainstream, but remains non-mainstream, like Celtic Spirituality, Hildegard of Bingen, St Francis, John Amos Comenius, etc.

14. To apply this threefold thought to Korean theology, Hoik Huh invented a new term, "Theocosmoandric Theology." For more about this subject, see http://mokja.info. For a critical review of this theological approach, see Soonkyung Park, *With the People: Those Who Remember the Past* (Seoul: Four Season, 2003), 238–46.

15. Cf. Andrew Callander, "Exploring a Christian Conception of Economic Life from within Karl Barth's Doctrine of Creation" (Ph.D. diss., University of Otago, Dunedin, New Zealand, 2012), 16–49.

16. Karl Barth, *Church Dogmatics: The Doctrine of Creation*, III/1 (Edinburgh: T&T Clark, 1982), 94–329.

17. Christoph Stückelberger, *Umwelt und Entwicklung: eine sozialethische Orientierung* (Stuttgart: W. Kohlhammer, 1997), 294–338.

18. Translation from the book *God Is Rice: Asian Culture and Christian Faith* (Geneva: WCC, 1986), slightly changed by Meehyun Chung. There is another version: Kim Chi Ha, *The Gold-Crowned Jesus and Other Writings,* ed. C. S. Kim (Maryknoll, NY: Orbis, 1978), 30.

19. This part is slightly changed and based on following article: Meehyun Chung, "Water and Rice: A Korean Eco-Feminist Theological Perspective," in *Breaking Silence: Theology from Asian Women*, ed. Meehyun Chung (Delhi: ISPCK, 2006), 90–102.

20. Youlku Kim, "Rice and Life," in *Talk about Rice*, ed. Newspaper of farmers (Seoul: n.p., 2005), 12.

21. Indica accounts for 90 percent of the world's rice production. Japonica is produced only in smaller areas like North Asia (including Korea), California (USA), North and South Australia, Northeast Africa, and Italy in Europe.

22. Sunghun Kim, "Ecological Economy of Rice," in *Talk about Rice*, 22.

23. The Hansalim Movement in Wonjoo since 1980 provides one important example of this.

24. Already in 1962 Rachel Carson mentioned the danger of chemicals and the pollution of water in her book *Silent Spring*. The chemical pollution of both the soil and water have become critical issues in the Korean peninsula as well, because organic farming is much more expensive and labour intensive. In order to compete in the global market, cheap production methods must be used, which involves spraying with insecticides. If rice farming is devalued and only industrialization is valued, the present decline of farming could escalate and our agricultural economy in the countryside could collapse under the pressure of the open market.

25. Historically the Demilitarized Zone (DMZ) was created in 1953 as part of the Korean Armistice Agreement between the People's Republic of China, North Korea, and the United Nations Command along approximately the 38th degree of latitude. Since nobody has lived there for more than 60 years, this area along the length of 250 km has become an unofficial nature reserve for biodiversity. The DMZ divides Korea roughly into two.

26. There are many attempts to make this issue more visible and to let victims be protected by law for compensation. For more about NGO efforts regarding this issue, see http://www.psakorea.org.

27. Cf. Maria Mies and Vandana Shiva, *Ecofeminism* (Halifax, NS: Fernwood, 1993), 13; Hyun Shik Jun, "Rosemary Ruether`s Critical Methodology and its Embodiment into Feminism and Ecology," *Korea Journal of Christian Studies* 48 (2006): 5–26.

28. For instance, the core value of Taoism, which is one essential part of indigenized Korean traditional thought and mind, is long life with blessing. Taoism escapes the round of daily anxiety by focusing on the order of the universe. To eat well, to live a healthy life, to have a long life with material blessings and eternal life are the basic values of Taoism, which have influenced Korean Christianity. This in turn reduces the biblical understanding of blessing to material blessings

for individuals and families. The Taoist idea of eternal life and the idea of a Taoist hermit with supernatural powers are confused with the Christian understanding of resurrection.

29. Ivone Gebara, *Longing for Running Water: Ecofeminism and Liberation* (Minneapolis: Fortress Press, 1999), 8.

30. Cf. Jooseop Keum, ed., *Together towards Life. Mission and Evangelism in Changing Landscapes* (Geneva: World Council of Churches Publications, 2013).

31. Cf. Alistair Kee, *Constantine versus Christ: The Triumph of Ideology* (Harrisburg, PA: Trinity Press International, 1982).

32. Keum, ed., *Together towards Life*, 5.

33. Ibid., 11.

34. Ibid.

35. Ibid., 38.

Chapter 16: Divest and Reinvest Now!

1. See "Presbyterian Church (USA) Divestment from Israel controversy," http://en.wikipedia.org/wiki/Presbyterian_Church_(USA)_disinvestment_from_Israel_controversy.

2. See Justin Gillis, "To Stop Climate Change, Students Aim at College Portfolios," *The New York Times*, 4 December 2012, http://www.nytimes.com/2012/12/05/business/energy-environment/to-fight-climate-change-college-students-take-aim-at-the-endowment-portfolio.html?pagewanted=all&_r=2&.

3. See Anil Markandya and Paul Wilkinson, "Electricity generation and health," *The Lancet* 370, no. 9591 (15 September 2007): 979–90, http://www.thelancet.com/journals/lancet/article/PIIS0140-6736(07)61253-7/fulltext.

4. Ibid.

5. See International Energy Agency, "Renewable Energy," http://www.iea.org/aboutus/faqs/renewableenergy/. See also Tim McDonnell, "Charts: The Smart Money Is on Renewable Energy," Mother Jones, 22 April 2013, http://www.motherjones.com/blue-marble/2013/04/charts-renewable-energy-fossil-fuels.

6. See Brad Plumer, "Why $775 billion in fossil-fuel subsidies are so hard to scrap," *The Washington Post*, 18 June 2012, http://www.washingtonpost.com/blogs/wonkblog/post/why-775-billion-in-fossil-fuel-subsidies-are-hardto-scrap/2012/06/18/gJQABaQUlV_blog.html.

7. See Steve Gelsi, "Fossil fuels get billions more in subsidies than renewables: IEA," *Market Watch*, 12 November 2012, http://blogs.marketwatch.com/thetell/2012/11/12/fossil-fuels-get-billions-more-in-subsidies-than-renewables-iea/.

8. See John Dominic Crossan *The Dark Interval: Towards a Theology of Story* (Santa Rosa, CA: Polebridge Press/Eagle Books, 2004).

9. Yanosek Cassia, "Policies for Financing the Energy Transition," in *Daedalus—Journal of the American Academy of Arts and Sciences* (Spring 2012), 94ff.

10. http://www.opensecrets.org/industries/indus.php?Ind=E01.

11. Jessica Leber, "Are Oil Companies Wasting Billions on Energy They'll Never Use?," *Fast Company*, http://www.fastcoexist.com/3020656/are-oil-companies-wasting-billions-on-energy-theyll-never-use. Frankfurt School/FS-UNEP Collaborating Centre for Climate & Sustainable Energy Finance, "Global Trends in Renewable Energy Investment," http://fs-unep-centre.org/publications/global-trends-renewable-energy-investment-2013.

12. http://kairos.technorhetoric.net/layers/metaphor.html.

13. As a member of the World Union for Progressive Judaism, the URJ connects Reform Jews in North America with Liberal/Progressive/Reform congregations around the globe. The URJ also represents Reform congregations in regional, North American and international organizations that include the Conference of Presidents of Major American Jewish Organizations and the Jewish Council for Public Affairs. For more information, see http://www.urj.org.

14. Union for Reformed Judaism, "South Africa," 58th General Assembly, November 1985, Los Angeles, California, http://urj.org//about/union/governance/reso//?syspage=article&item_id=2096.

15. At times, the refusal of governments to halt oppressive policies, such as apartheid in South Africa, has led religious groups to divest from major companies doing business with the government or in its country.

16. See, for example, "Create Climate Future Taks Force, 2013—Exxon Mobil Corporation," http://www.onlineethicalinvestor.org/eidb/wc.dll?eidbproc~reso~10279.

17. Desmond Tutu, "Divesting from Injustice," *The Huffington Post*, 13 June 2010, http://www.huffingtonpost.com/desmond-tutu/divesting-from-injustice_b_534994.html.

18. Kate Galbraith, "Churches Go Green by Shedding Fossil Fuel Holdings," *The New York Times*, 15 October 2014, http://www.nytimes.com/2014/10/16/business/international/churches-go-green-by-shedding-holdings-of-carbon-emitters.html.

19. For a link to this story, and for a comprehensive listing of other fossil-fuel divestment resolutions, see GreenFaith, "A Listing of Known Religious Divest and Reinvest Efforts," http://www.greenfaith.org/programs/divest-and-reinvest/listing-of-known-religious-divestment-efforts/.

20. Uniting Earth Web, "Go fossil free," http://www.unitingearthweb.org.au/fossil-fuels/31-go-fossil-free.html.

21. Micki Carter, "United Church of Christ to become first U.S. denomination to move toward divestment from fossil fuel companies," United Church of Christ, 30 June 2013, http://www.ucc.org/news/GS2013-fossil-fuel-divestment-vote.html.

22. Anne van Staveren, "Quakers to divest from fossil fuels," Quakers in Britain, 8 October 2013, http://www.quaker.org.uk/news/quakers-disinvest-fossil-fuels.

23. Aaron Packard, "Success! 5 Key Factors in the Anglican Church of NZ & Polynesia's Vote to Divest," 350.org, 19 May 2014, http://350.org/success-5-key-factors-in-the-anglican-church-of-nz-polynesias-vote-to-divest/.

24. "Fossil Fuel Divestment, Unitarian Universalist Association 2014 Business Resolution," http://divestfossilfuels.files.wordpress.com/2013/11/fossil-fuel-divestment-resolution.pdf.

25. "Union Theological Seminary in the City of New York Votes Unanimously to Divest from Fossil Fuels," Union Theological Seminary, 10 June 2014, http://www.utsnyc.edu/divestment.

26. Serene Jones, "Union Becomes the World's First Seminary to Divest from Fossil Fuels," *Time*, 10 June 2014, http://time.com/2853203/union-fossil-fuels/.

27. See "Dayton Divests," University of Dayton, 23 June 2014, https://www.udayton.edu/news/articles/2014/06/dayton_divests_fossil_fuels.php.

28. Ibid.

29. GreenFaith, "Divestment in Faith-Based Judicatories, International," http://greenfaith.org/programs/divest-and-reinvest/listing-of-known-religious-divestment-efforts#divestment-in-faith-based.

30. Adam Vaughn, "World Council of Churches rules out fossil fuel investments," *The Guardian*, 11 July 2014, http://www.theguardian.com/environment/2014/jul/11/world-council-of-churches-pulls-fossil-fuel-investments.

31. "World Council of Churches Endorses Fossil Fuel Divestment," 350.org, 11 July 2014, http://350.org/press-release/world-council-of-churches-endorses-fossil-fuel-divestment/.

32. Melanie Mattauch, "Oxford Diocese goes fossil free and challenges Church of England to divest," Fossil Free, 15 November 2014, http://gofossilfree.org/oxford-diocese-goes-fossil-free-and-challenges-church-of-england-to-divest-2/.

33. See the Sustainable Endowments Coalition, *Greening the Bottom Line 2012*, http://greenbillion.org/wp-content/uploads/2012/11/Greening-the-Bottom-Line-2012.pdf.

Chapter 17: Indigenous Ecological Spirituality and the Land

1. Northeast India is a region poorly connected to the Indian mainland by a small corridor, and is surrounded by Bhutan, Myanmar, Bangladesh, and China. The traditional tribes in the Northeast are historically linked as they are largely of Tibeto-Burman/Mongoloid stock, and are closer to Southeast Asia than to South Asia. Their region is ethnically, linguistically, and culturally distinct from the other states of India. Interestingly, 99 percent of the Northeast's boundaries are international and only 1 percent is domestic. The eight states comprising the Northeast are populated by nearly 40 million inhabitants who among themselves vary in language, race, tribe, religion, and regional heritage. Therefore, most often, the clubbing of all these states under the tag of "Northeast" has tended to have a homogenizing effect, with its own set of implications for policy formulation and implementation, not to mention local aversion to such a construct.

2. A. P. J. Abdul Kalam with Y. S. Rajan, *India 2020: A Vision for the New Millennium* (New Delhi: Penguin Books, 2002), 12.

3. Himanshi Dhawan, "Northeast emerges as the new hotspot for tourists," *The Times of India*, 11 August 2013, http://timesofindia.indiatimes.com/business/india-business/Northeast-emerges-as-the-new-hotspot-for-tourists/articleshow/21753318.cms.

4. "North Eastern Region," ch. 7 in Planning Commission, Government of India, *Eleventh Five Year Plan (2007–2012)*, vol. 1, *Inclusive Growth* (New Delhi: Oxford University Press, 2008), 167, http://planningcommission.nic.in/plans/planrel/fiveyr/11th/11_v1/11th_vol1.pdf.

5. India's LEP, launched in 1992 with the objective of fostering closer ties with Southeast Asian nations, is one of the most important cornerstones of India's foreign policy. Apart from the obvious economic benefits, India also wants to send an unequivocal message to China of its aim to become a key regional power, which will be achieved by forging closer ties with countries in Southeast Asia to counter China's increasing dominance in the region. To this end, strong advances have been made toward establishing closer ties with Thailand, Myanmar, and Vietnam.

6. For more details see "North Eastern Region," in Planning Commission, *Eleventh Five Year Plan*, 168–75, http://planningcommission.nic.in/plans/planrel/fiveyr/11th/11_v1/11th_vol1.pdf.

7. Pushpita Sen's report on the "Roundtable on Northeast's Role in India's Look East Policy," a half-day programme that took place on 18 September 2013 and was organized by the Institute for Defence Studies and Analyses (IDSA), with the purpose of inviting specific projects from the northeastern states that could be implemented in the region under the aegis of the ASEAN-India Vision Document 2020 as well as India's LEP. See http://e-pao.net/epSubPageExtractor.asp?src=education.Scientific_Papers.Roundtable_on_Northeast_role_in_India_Look_East_Policy.

8. Suggestions have also been made that in order to create economic interests for Myanmar in India and encourage it to act against the northeast insurgent groups and keep the road networks secure, people in the Northeast could be induced to take up land on lease for cultivation for various crops in Myanmar and set up food-processing units in Myanmar. For more details, see ibid.

9. The earmarked amount was about 10 percent of the ministry's total allocation, though the actual release for the region has been much higher during the 11th Plan. The central budget provisioning for the NEI region has also gone up in recent years. Total outlay for the region stood at Rs. 88 crores (USD 14.7 million) and Rs. 93 crores (USD 15.5 million) respectively for product/infrastructure development in identified destinations and circuits during 2010–11 and 2011–12, representing around 16 percent of central allocation under this head. See *Draft: National Tourism Policy of India*, 20, www.sihra.org/downloads/DraftNTP.doc.

10. Vidya Rangan, "Paradise Unexplored—At What Price?," in *IFIs and Tourism: Perspectives and Debates*, Equations (March 2008), 52.

11. Describing himself as the "adopted son of the North-East region" (he is a Rajya Sabha/Upper House member from Assam), Manmohan said Vision 2020 will pave the way for the region's future development. Quoted by Anil Anand,

"Manmohan unveils vision 2020 for North East," *DNA*, http://www.dnaindia.com/india/1175232/report-manmohan-unveils-vision-2020-for-north-east.

12. For more details see Ministry of Development of North Eastern Region and North Eastern Council, Government of India, *North Eastern Vision of 2020*, 3, 18, 28, http://www.mdoner.gov.in/content/vision-document.

13. Some of the things NEI is considered to offer are tribal lifestyle, heritage, tea garden, golf, monasteries at the one end and eco-tourism, adventure, and wild life on the other. Ibid., 127, 128.

14. Ibid., 119.

15. Ibid., 129, 131.

16. Ibid., 133.

17. See "The Proposal of Promoting the Northeast as Destination to Leave Travel Concession (LTC) Travelers," Equations, 12 June 2008, http://www.equitabletourism.org/campaign.php?AID=338.

18. This has already been happening in other parts of India, like Bekal, Mahabalipuram, Kovalam, and Sindhudurg, identified as Special Tourism Areas (STAs) in the 1992 National Tourism Policy, which have turned into enclaves of investment, exploitation and isolation from their surroundings. See Rangan, "Paradise Unexplored," 62–93.

19. The notion of sustainable tourism is intrinsically rooted in the Western concept of environmentalism, which does not take into account basic livelihoods needs of communities based in poor regions of Asia and therefore cannot be adapted to situations of chronic poverty, glaring inequities, and even conflict-prone zones. See ibid., 58.

20. Eco-tourism is widely regarded as the ideal manner to promote tourism with little or no damage to the environment. Its proponents disagree that it can be synonymous with tourism development in general because this form of tourism is low impact and environmentally sustainable.

21. Consultation on the "Churches' Response to Development and Tourism," jointly organized by the Ecumenical Coalition on Tourism (ECOT) with National Council of Churches in India (NCCI), Presbyterian Churches in India (PCI), and Martin Luther Christian University (MLCU), a Northeast India regional consultation from 29–31 March 2011 at PCI, Assembly Hall, in Shillong, Meghalaya, India.

22. Emphasis was made on preserving water for the lean period and to help tourism to flourish as an industry in the state. The Department of Tourism, Government of Nagaland and the Nagaland Tourism Association (NTA) came together in organizing the main function at the Tourism Directorate Kohima under the theme "Tourism & Water: Protecting our common future." Speaking on the occasion, P. Angami, head of the tourism department, called upon the stakeholders of tourism activities in the state to come together in ensuring sustainable access to water resources in their own small ways. NTA president V. Vincent Belho stressed the need to preserve water so that tourism can flourish in the state. He said tourism is an industry and efforts should be made toward promoting and making the state a tourist hot spot. He also called upon tour operators to help in improving the economy of the people of the

state down to the village levels and also asked them to ensure environmental sustainability. See "World Tourism Day observed in Nagaland," *The Times of India*, 27 September 2013, http://timesofindia.indiatimes.com/city/guwahati/World-Tourism-Day-observed-in-Nagaland/articleshow/23160456.cms.

23. Studies of sacred groves in Manipur have revealed that therapeutic application of 127 plant species have been discovered in its sacred groves, which are protected by the local people based on the grounds of indigenous culture and religions. The people inhabiting the areas near the sacred groves explain how the knowledge about the medicinal plants learned from medicine men known as *Maibas* is being passed down from one generation to the other. See Ashalata Devi Khumbongmayum, M. L. Khan, and R. S. Tripathi, "Ethnomedicinal plants in the sacred groves of Manipur," in *Indian Journal of Traditional Knowledge* 4, no. 1 (January 2005): 21–32, at 21, http://nopr.niscair.res.in/bitstream/123456789/8490/1/IJTK%20 4%281%29%2021-32.pdf.

24. *Who Really Benefits from Tourism? Documentation Updates: January–September 2004*, Equations, 57; http://www.scribd.com/doc/30497929/ Documentation-Update-January-to-September-2004#scribd.

25. The only type of employment open to the local people is at the bottom rung of the job ladder. The local people are to be employed as porters, cooks, cleaners, gardeners, drivers, security guards, entertainers, and so on. They are to be paid minimum wages. Even these menial jobs often lack any formal contracts and are subject to the tourist season.

26. This idea has been put forward in Bryan H. Farrell and Dean Runyan, "Ecology and Tourism," *Annals of Tourism Research* 18, no. 1 (1991): 26–40, at 31, http:// citeseerx.ist.psu.edu/viewdoc/download?doi=10.1.1.471.2171&rep=rep1&type=pdf.

Chaper 18: The Dance of Krishna

1. A. C. Bhaktivedanta Swami, *Krsna: The Supreme Personality of Godhead* (Alachua, FL: Bhaktivedanta Book Trust, 1970).

2. Steven C. Sherwood et al., "Spread in model climate sensitivity traced to atmospheric convective mixing," *Nature* 505 (2 January 2014): 37–42.

3. Prabhakar Nair, "The Coastal Eco Region—Vulnerabilities of Fishing Communities to Ecological and Climate Change," in *Voices of Vulnerability, Choices of Change*, ed. John D'Souza (Andhra Pradesh: INECC, 2012), 53.

4. E. Vivekanandan, "Impact of Climate Change in the Indian Marine Fisheries and the Potential Adaptation Options," in *Coastal Fisheries Resources of India—Conservation and Sustainable Utilisation*, ed. B. Meenakumari et al. (Cochin: Society of Fisheries Technologists-India, 2010), 169–84.

5. E. Vivekanandan, "Impact and Adaptation Options for Indian Marine Fisheries to Climate Change," *Central Marine Fisheries Research Institute*, 65, http://eprints.cmfri.org.in/9832/1/ICT_Oriented_Strategic_Extension_for_Responsible_Fisheries_Management_Manual-87-96.pdf.

6. Ibid.

7. Derek Staples and Simon Funge-Smith, *Ecosystem Approach to Fisheries and Aquaculture: Implementing the FAO Code of Conduct for Responsible Fisheries* (Bangkok: Food and Agriculture Organization of the United Nations Regional Office for Asia and the Pacific, 2009), 3, ftp://ftp.fao.org/docrep/fao/012/i0964e/i0964e00.pdf.

8. M. Thrane, "LCA of Danish Fish Products—New Methods and Insights," *International Journal of Life Cycle Assessment* 11 (2006): 66–74, at 69, cited in Vivekanandan, "Impact and Adaptation Options" (see n. 5, above).

9. Food and Agriculture Organisation (FAO), "Building Adaptive Capacity to Climate Change: Policies to Sustain livelihoods and Fisheries," FAO *Policy Brief* 8:16, cited in ibid.

10. Prabhakar Nair, "Vulnerabilities of Fishing Communities to Ecological and Climate Change," in D'Souza, ed., *Voices of Vulnerability*, 56.

11. Anupam Chakravartty, "Expect more extreme climate events in India," *Down to Earth* (31 March 2014), http://www.downtoearth.org.in/content/expect-more-extreme-climate-events-india.

12. Prabhakar Nair, "Vulnerabilities of Fishing Communities to Ecological and Climate Change" in D'Souza, ed., *Voices of Vulnerability*, 72.

13. Chakravartty, "Expect More Extreme Climate Events in India."

14. Ashish Agarwal, et al., *Is India Ready to Implement REDD PLUS? A Preliminary Assessment*, (New Delhi: The Energy and Resources Institute [TERI]: 2009), http://www.teriin.org/events/CoP15/Forests.pdf.

15. Sanjay Khatua, DHARA, "The Forest Eco-region—Vulnerability of the Forest Eco ystem in Context of the Changing Climate," in D'Souza, ed., *Voices of Vulnerability*, 81.

16. Ranchor Prime, *Hinduism and Ecology: Seeds of Truth* (New Delhi: Motilal Banarasidass, 1994), 92.

17. "Forests and Climate Change," Forests Forever, http://www.forestsforever.org/climate2.html.

18. Ashish Aggarwal, "Implementation of Forest Rights Act, Changing Forest Landscape, and 'politics of REDD+' in India," *Resources, Energy, and Development* 8, no. 2 (September 2011): 131–48, http://www.teriin.org/projects/nfa/cc2bwp2.php.

19. "By 2050, 54% of India Population to Shift to Urbanised Cities," Commodity Online, 7 April 2012, http://www.commodityonline.com/news/by-2050-54-of-india-population-to-shift-to-urbanised-cities-47263-3-47264.html.

20. Bhavika Jain, "62% of Mumbai Lives in Slums: Census," *Hindustan Times*, 17 October 2010, http://news.inmumbai.org/2010/10/17/62-of-mumbai-lives-in-slums-census.

21. Dipak Kumar Dash, "By 2017, India's Slum Population Will Rise to 104 Million," *Times of India*, 20 August 2013, http://timesofindia.indiatimes.com/india/By-2017-Indias-slum-population-will-rise-to-104-million/articleshow/21927474.cms.

22. Sayantan Bera, "Kolkata Is One of World's 10 Cities Most at Risk from Natural Disasters," Down to Earth, 27 March 2014, http://www.downtoearth.org.in/content/kolkata-one-world-s-10-cities-most-risk-natural-disasters.

23. Architesh Panda, "Climate Change Risks and Adaptation: Indian Mega Cities," *The India Economy Review*, 1 January 2011, http://theindiaeconomyreview.org/Article.aspx?aid=30&mid=3.

24. Vinayashree Jagadeesh, "City's Air Dangerously Polluted," *The Hindu*, 25 August 2014, http://www.thehindu.com/todays-paper/citys-air-dangerously-polluted/article6348230.ece.

25. Translation of Shrimad Bhagavatam 10.21.18, http://vedabase.net/sb/10/21/18/.

26. Martin Beniston and Douglas G. Fox, "Impacts of Climate Change on Mountains," in *Climate Change 2014: Impacts, Adaptation, and Vulnerability*, IPCC AR5 (March 2014): 194, https://ipcc-wg2.gov/AR5/images/uploads/WGIIAR5-PartA_FINAL.pdf.

27. Mukesh Ray et al., "Mountain Eco-region-Climate Vulnerability in North Western Himalayas," in D'Souza, ed., *Voices of Vulnerability*, 15.

28. Javaid Laghari, "Climate Change: Melting Glaciers Bring Energy Uncertainty," *Nature: International Weekly Journal of Science* (15 November 2013), http://www.nature.com/news/climate-change-melting-glaciers-bring-energy-uncertainty-1.14031.

29. "Uttarakhand Flash Flood—A Report," Indian Red Cross Society (27 June 2013), http://www.indianredcross.org/press-rel27-june2013.htm.

30. Poonam K. Singh and Ramesh C. Dhiman, "Climate Change and Human Health: Indian Context," *Journal of Vector Borne Diseases* 49 (June 2012): 55–60, http://www.mrcindia.org/journal/issues/492055.pdf.

31. J. P. Majra and A. Gur, "Climate Change and Health: Why Should India Be Concerned?," *Indian Journal of Occupational and Environmental Medicine* 13, no. 1 (April 2009): 11–16, http://www.ncbi.nlm.nih.gov/pmc/articles/PMC2822161/.

32. Ibid.

33. Singh and Dhiman, "Climate Change and Human Health."

34. Dileep Mavalankar et al., "Climate Change and Health Preparedness in India: Protecting Local Communities in Ahmedabad, Gujarat from Extreme Heat," *Natural Resources Defense Council* (2012): 8, http://www.nrdc.org/international/india/files/IndiaHealthReport.pdf.

35. Valerie J. Roebuck, trans., *The Upanishads* (New Delhi: Penguin India, 2000), 58.

36. "Warming Climate in India to Pose Significant Risk to Agriculture, Water Resources, Health, says World Bank Report," The World Bank Press Release (19 June 2013), http://www.worldbank.org/en/news/press-release/2013/06/19/warming-climate-india-pose-significant-risk-agriculture-water-resources-health-says-world-bank-report.

37. Chakravartty, "Expect More Extreme Climate Events in India."

38. Rahmat Tarikere, "Food et cetera," India Seminary (2010), http://www.india-seminar.com/2010/612/612_rahmat_tarikere.htm.

39. "Children in India—A Statistical Appraisal," Social Statistics Division, Ministry of Statistics and Programme Implementation, Government of India (September 2012), http://mospi.nic.in/mospi_new/upload/Children_in_India_2012.pdf.

40. Dhan Foundation, "Supporting Millets in India: Policy Review and Suggestions for Action," Revalorising Small Millets in Rainfed Regions of South Asia (RESMISA) (August 2012), 2, http://www.dhan.org/smallmillets/docs/report/Millet_Support_Policies.pdf.

41. Ibid., 11.

42. Ibid., 28.

43. "Talk: Albert Einstein," http://en.wikiquote.org/wiki/Talk:Albert_Einstein.

44. "Hindu Declaration on Climate Change," Presented for Consideration to the Convocation of Hindu Spiritual Leaders Parliament of the World's Religions, Melbourne, Australia, 8 December 2009, http://www.hinduismtoday.com/pdf_downloads/hindu-climate-change-declaration.pdf.

45. Roebuck, *The Upanishads*, 3.

46. Prime, *Hinduism and Ecology*, 115.

Epilogue: Theological Reflections on Ecumenical Action and Advocacy on Climate Change

1. For an earlier contribution in this regard, see David G. Hallman, "Climate Change and Ecumenical Work for Sustainable Community" in *Earth Habitat: Eco-injustices and the Church's Response* ed. Dieter T. Hessel and Larry L. Rasmussen (Minneapolis: Fortress Press, 2001), 125-134.

2. See the prophetic booklet entitled *Solidarity with the Victims of Climate Change: Reflections on the World Council of Churches' Response to Climate Change* produced by the WCC for the World Summit on Sustainable Development in Johannesburg in 2002 (Geneva: WCC, 2002).

3. See Ernst M. Conradie, "Confessing Guilt in the Context of Climate Change: Some South African Perspectives" in *Ecological Awareness: Exploring Religion, Ethics and Aesthetics* ed. Sigurd Bergmann and Heather Eaton (Berlin: LIT Verlag, 2011), pp. 77-96.

4. See http://earthstatement.org/statement/ (accessed 6 June 2015).

5. See http://www.globalcarbonproject.org/carbonbudget/15/hl-compact.htm (accessed 29 February 2016).

6. Katja Frieler (Potsdam Institute for Climate Impact Research) reports that "At the Paris climate summit, delegates have struck an agreement that calls for the world to "hold the increase in the global average temperature to well below 2° above pre-industrial levels and to pursue efforts to limit the temperature increase to 1.5°. But the climate action pledges made by 185 countries ahead of the summit don't add up to 1.5° or warming or even 2°. Taken together, they add up to a 2.7° world. As the negotiations go on, 2015 is about to set a new global temperature record, and is likely to have reached 1° warming already. See http://theconversation.com/this-is-what-will-happen-to-the-climate-in-the-next-100-years-52051 (accessed 29 February 2016).

7. As reported in *The Guardian*. See http://www.theguardian.com/environment/2015/apr/22/earth-day-scientists-warning-fossil-fuels- (accessed 6 June 2015).

8. See also Ernst M. Conradie, "Appropriate Contemporary Forms of Apocalyptic," in Hille Haker, André Torees Queiruga & Marie-Theres Wacker (eds), *The Return of Apocalypticism*, *Concilium* 2014/3, 96-103.

9. See, most recently, Ernst M. Conradie, *The Earth in God's Economy*: *Creation, Salvation and Consummation in Ecological Perspective.* Berlin: LIT Verlag, 261.

10. See, for example, James Lovelock's *The Vanishing Face of Gaia: A Final Warning* (New York: Basic Books, 2009).

11. See http://earthstatement.org/statement/ (accessed 6 June 2015).

12. See *Solidarity with the Victims of Climate Change*, p. 25-26.